Peter Loughran, Bruce Coleman Ltd.

HIDDEN WORLDS

FROM BUD TO BLOSSOM, *a hibiscus flower opens before your eyes (above). You would have to be very patient to see this happen in nature. The process usually takes several days. To capture the action, the photographer made pictures of the flower on different days. Each picture was made from exactly the same spot and angle. The result: a flower that "blooms" as you watch.*

Would you like to have superhuman eyes? This book will give them to you. You'll spy on tiny plants and animals that live in a drop of water. You'll explore distant stars, planets, and galaxies. You'll find out how birds and insects see things. This book will show you the world and the heavens as you've never seen them before.

Library of Congress ℂℙ data: page 104

BOOKS FOR WORLD EXPLORERS
NATIONAL GEOGRAPHIC SOCIETY

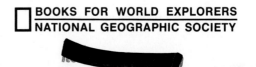

CONTENTS

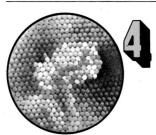

2   WHILE PLAYING OUTSIDE, *these children see a lot of things, but they can't see everything around them (right). The small circles reveal some of the things they are missing—things you will see in this book.*

Robert E. Hynes

1 THINK SMALL

by Edith Kay Pendleton

magine that you have found a pair of magic glasses. When you put them on, tiny things suddenly seem large. An inch-long mayfly looks as big as a small bird (right). The threads in a nylon stocking seem to be the size of links in a metal fence. Grains of sugar resemble precious gems.

The people who discovered magnifying lenses might have thought they were magic. At first, they could see things only a few times larger than life-size. Now we have instruments so powerful they can magnify tiny portions of objects *two million times!*

Turn the page and enter the hidden world of magnification, a world where small becomes large.

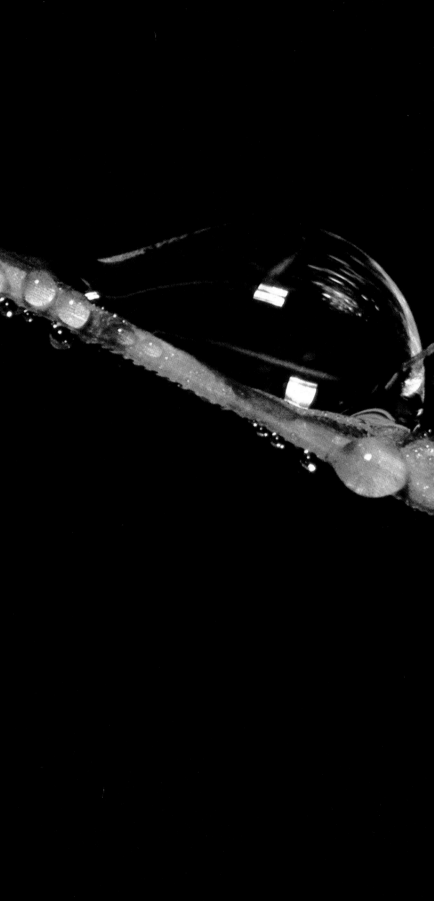

SPARKLING DROPS OF WATER *cling to a mayfly and a leaf, both magnified by a special camera lens. Mayflies spend most of their lives in ponds, as swimming creatures. Then, like mosquitoes, they change into adults with wings. Every spring, millions leave the water. They rest on plants and shed their skins. Within 24 hours they mate in the air, lay eggs, and die. The eggs fall into the water, and the cycle of life begins once more.*

4 Hans Pfletschinger from Peter Arnold, Inc.

ZOOMING IN

You know what an orange looks like. Or do you? A close look shows that the skin has tiny pits in it (right). Greatly magnified (far right), the skin becomes a rocky landscape. There's more to an orange peel than meets the eye!

OPENING A HIDDEN WORLD, *Kevin Walker, 11, of Washington, D. C., starts to peel an orange (above). Thousands of tiny bags lie just below the surface of the skin. As Kevin tears the skin, the bags burst, spraying mists of fragrant moisture and oil. The next time you peel an orange, sniff your hands afterward. You may smell oil clinging to them.*

UNDER THE SKIN, *Kevin finds a layer of white, spongy material (right). While the fruit is growing on the tree, this layer helps protect it from the cold.*

6

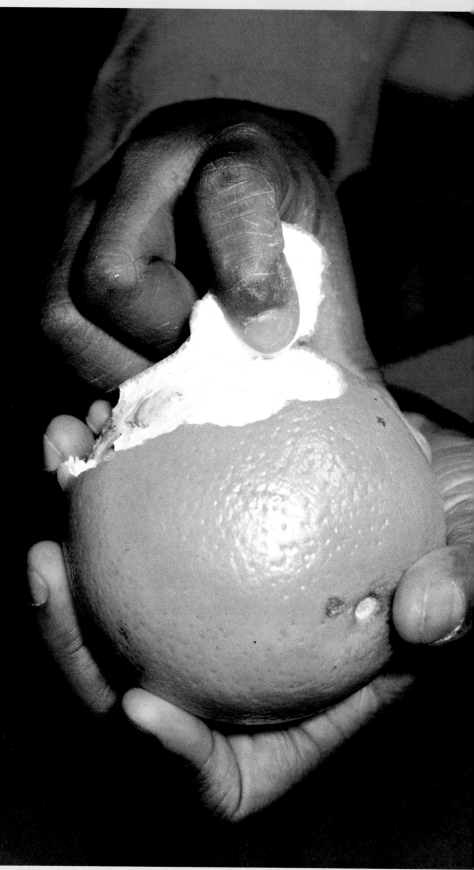

National Geographic Photographer Joseph D. Lavenburg (both)

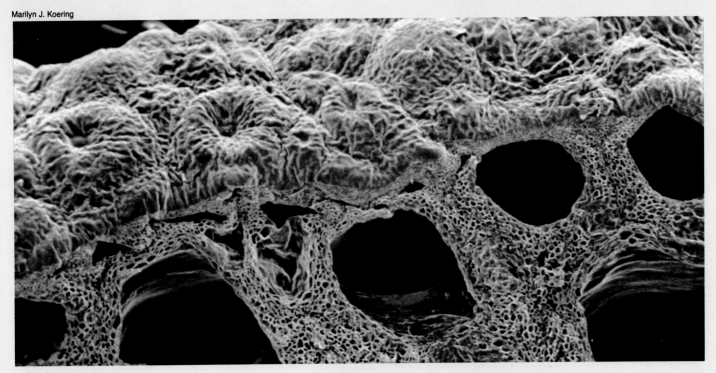

SMOOTH SKIN *looks rough when a dried-out orange peel is highly magnified. The empty oil bags look like dark caves.*

SEE BIG. *A hand-held magnifying glass is the simplest kind of microscope. The glass is ground into a curved shape, then fitted into a frame. The curved glass bends rays of light as they pass through it. This makes things seem larger than they are. Most hand lenses double the size of objects. People began making magnifying glasses in the 1200s. They used them for reading.*

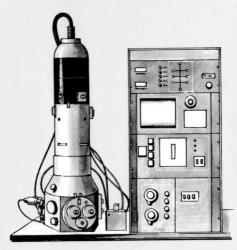

SEE BIGGER. *A compound microscope has two or more curved lenses in a tube. The object to be studied is placed on glass so that light from underneath passes through it. The first lens enlarges the image. Other lenses enlarge the enlargement. The observer may see things from 10 to 2,000 times normal size. The first microscopes were made around 1600.*

SEE BIGGEST. *A modern magnifier, the scanning electron microscope (SEM) has no lenses. Beams of tiny particles called electrons bounce off an object inside the SEM cylinder. They form an enlarged three-dimensional image. It appears on a screen, like a TV picture. The picture is always in black and white. Sometimes colors are added later. You will see SEM photographs on pages 19-25. The SEM was developed in the 1960s. It can magnify a small object, such as a virus, up to 600,000 times life-size. Some more complicated electron microscopes enlarge things millions of times. But they magnify only a very thin slice of the object.*

7

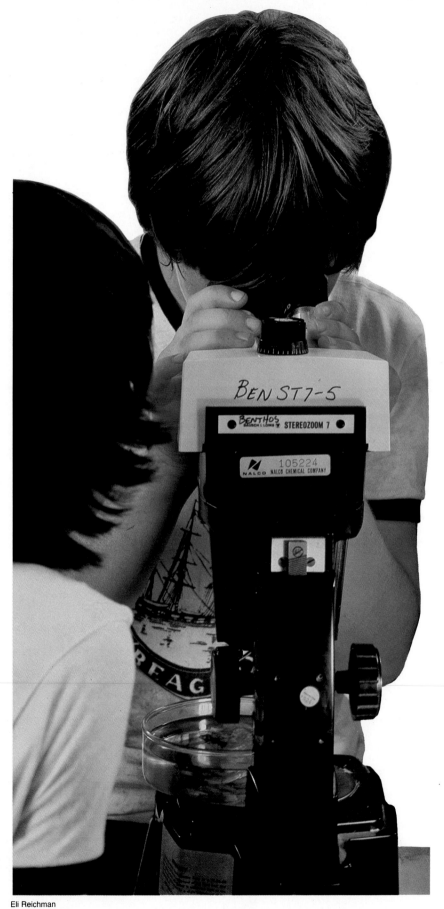

WHAT'S GOING ON DOWN THERE?
Peering through a microscope (left), a camper studies a small spider crab. It was brought up in the Oceanic's net. The camper uses a microscope that enlarges an entire creature—not just a thin slice preserved on a glass slide. Below, you can see what the camper sees.

MONSTER FROM THE DEEP? *No, it's just the head of a spider crab, magnified. The crab puts bits of plants and sponges on its shell. These help disguise it and hide it from enemies.*

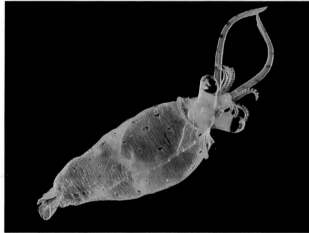

BABY SQUID *looks large through a microscope (above). Actually it is about half an inch (1 cm)* long. Young squids spend their first weeks of life near the surface. There they feed on small plants. As they mature, they spend the daylight hours on the ocean floor. They come to the surface to feed only at night.*

*Metric figures in this book are given in round numbers.   13

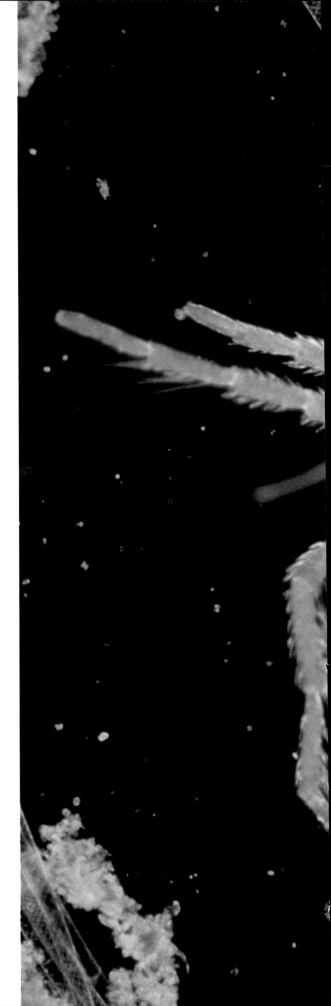

becomes a crew member with a special job to do. Some measure the depth of the water. Others prepare nets that will be used for collecting sea creatures.

In deeper waters, the net is let out. As the ship drags it along the bottom, the net gathers plants and animals. Everyone helps haul the net in. Campers pour the contents into water-filled trays. Then they sort their finds.

Some creatures are tossed back into the sea. Some are studied at once, in a laboratory on board the ship. Some go into water-filled tanks for future study on shore.

Whenever the ship is close to shore, campers scoop up water and examine it under powerful microscopes. They study bacteria and other tiny forms of life that live in shallow water.

At the end of a day, the campers return to shore with new knowledge about the sea and the hidden worlds it holds.

Kim Taylor, Bruce Coleman Ltd. (below and opposite)

INSPECTING ITS DINNER *with long feelers, a tiny creature called an amphipod (say AM-fih-pod) begins eating part of a dead plant. Amphipods are mostly scavengers. They eat dead plant and animal matter. The tiny creatures crawl about on the seafloor. There, sea plants help hide them. When they come out of their hiding places, they are often snapped up by hungry fish. Thirty or forty amphipods would fit on a penny. Amphipods are related to crabs, shrimps, and lobsters. All have soft bodies covered by hard shells. As they grow, they shed their shells and form new ones.*

WATER MITE *the size of a pinhead (right) lives in ponds and freshwater streams. It stays alive even when the water freezes. Legs covered with small hairs move the mite through the water as it looks for smaller creatures to eat. Its bright red coloring may warn enemies not to eat the mite because it tastes bad.*

14

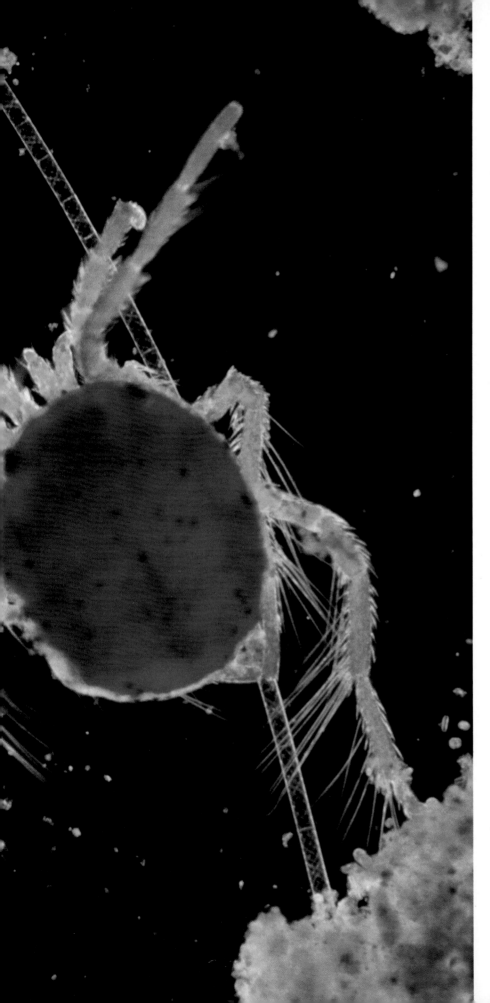

LIKE JEWELS AND RIBBONS, *tiny animals and plants float in a drop of water under the microscope (below). The animals are radiolarians (say ray-dee-oh-*LAR*-ee-uhns). Transparent skeletons, often with spikes, protect their soft, colorful bodies. The plants are algae (say* AL*-jee). Some kinds are round. Other kinds form long chains. A million algae can live in a teaspoonful of fresh or salt water. They provide food for many water creatures.*

Oxford Scientific Films

T. E. Adams from Peter Arnold, Inc.

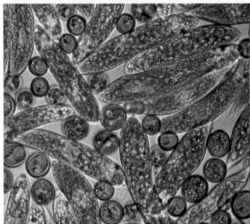

UNDER A MICROSCOPE, *pond water swarms with tiny creatures (above). In warm weather they multiply rapidly, forming surface scum. The long green creatures are swimmers called euglenas (say you-*GLEE*-nuhs). The brown ones are drifters called trachelomonads (say traa-kel-o-*MON*-ads).*

15

SEE INSIDE A STEM (right).
*A thin slice of a pond lily, dyed
to show up clearly under the
microscope, has been magnified
500 times. Cells in the stem
form many hollow tubes.
Together they resemble a bundle of
drinking straws. Most of the tubes carry
air from the leaves to the roots. Only the
large cell-filled tube in the center carries
food back and forth between the roots and
the leaves. The air-filled tubes help hold
the plant upright in the water.*

STEM →

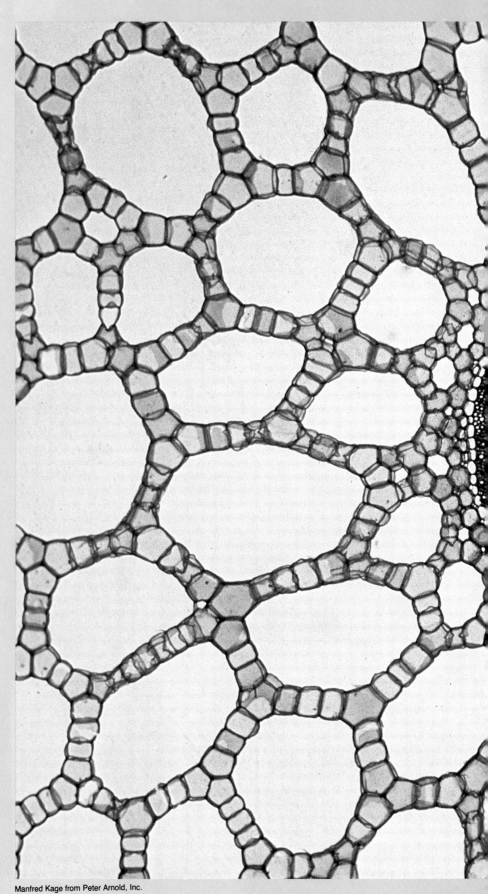

J. Robert Waaland

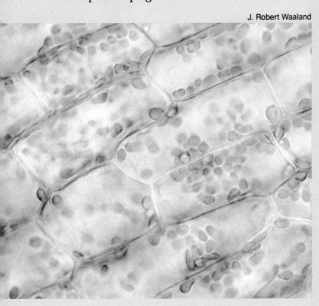

LOOK AT A LEAF (above).
*Under the microscope,
individual cells show up as
rectangles. This leaf belongs to
a water plant called an elodea
(say ih-LOHD-ee-uh). It is
found in many lakes. You may
have seen it in an aquarium. Each
rectangle is one cell in the leaf. The
darker green round shapes are chloroplasts
(say KLOR-uh-plasts). The green coloring
matter in chloroplasts absorbs sunlight and
uses it, along with material from water
and air, to make food for the plant.*

16

Manfred Kage from Peter Arnold, Inc.

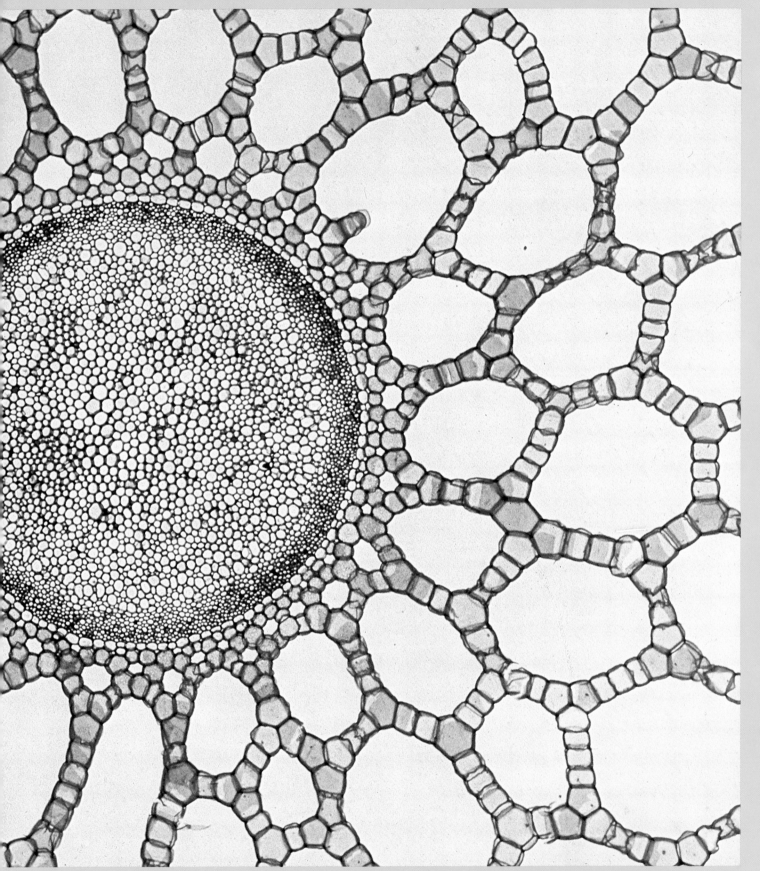

Hans Pfletschinger from Peter Arnold, Inc.

LOOK AROUND YOU

If you want to see strange sights in your own backyard, take a magnifying lens outside. Watch a bee as it feeds on a flower (above). Notice the pollen on its head, body, and legs. Look at its face. See how pollen clings to its antennae and to the fine hairs on its eyes (far right). A highly magnified view of a pollen grain (right) shows you one reason why. The pollen has a rough surface that easily catches on other rough objects—like hairy bees.

POLLEN OR PLANET? *Spikes and deep pits make a grain of dandelion pollen look like an alien planet (right). Each kind of pollen has its own shape. A powerful scanning electron microscope (*SEM*) made this picture. A yellow tint was added later. The tint gives the pollen natural coloring. The grain shown here has been magnified 3,000 times.*

18

HAIRY HEAD. *A hand lens reveals fine hairs on the eyes of a bee (below). They bend in the breeze and tell the bee whether it will have trouble flying in the wind. You can also see hairs on the bee's tongue. As the bee gathers nectar, other things stick to its tongue and face. Here, you see bits of the flower dangling from its mouth. The bee also picks up pollen. Some brushes off on other flowers. The pollen enables flowers to form seeds.*

19

MILLIONS OF SCALES cover a butterfly's wing. The SEM shows how they overlap (below). The log shape is a vein in the wing. Scales may be long or short, curly or stiff. They don't help a butterfly fly, but they do give the wings their coloring. Wing colors and patterns camouflage the insect and help it attract a mate. Scales can be knocked off easily. Bald spots show where some are missing on this wing. The scales don't grow back.

KEEPING CLEAN, a velvety tree ant wipes an antenna with one foot (right). The insect is only a quarter of an inch (two-thirds of a cm) long. This SEM portrait shows how the ant got its name. It is covered with soft hair that looks somewhat like velvet. Small, round eyes may sense only light and motion. The insect also collects information through its antennae. They detect scents, sounds, temperature changes, and movements.

Linda C. Sawyer

WORLD-CLASS JUMPER. An ordinary flea (right) can leap 50 to 100 times the length of its body. If you jumped that well, you could spring the length of a football field in one or two bounds! The flea has strong legs and sharp claws. It leaps onto a passing animal, hangs on with its claws, and digs in. Sharp mouth parts pierce the victim's skin and suck the blood. Instead of eyes, the flea has eyespots. It senses only light, which it avoids.

David Scharf from Peter Arnold, Inc. (above and opposite)

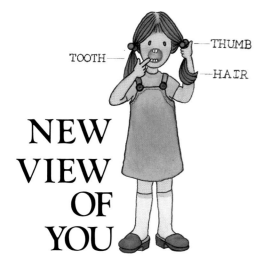

TOOTH — THUMB — HAIR

NEW VIEW OF YOU

FINGERTIP. A "cave" with rough walls looks spooky (right). Mossy "rocks" lie here and there. The "cave" is a pore, the opening of a sweat gland in your skin. Sweating helps keep you cool. Color has been added to this SEM enlargement. The green "rocks" are bacteria. They live everywhere on your skin. Most help you by eating disease-causing bacteria before they can do harm.

Manfred Kage from Peter Arnold, Inc.

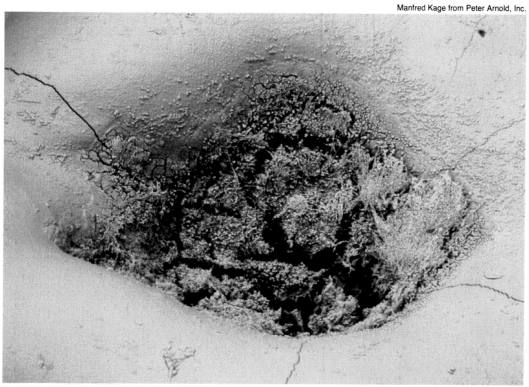

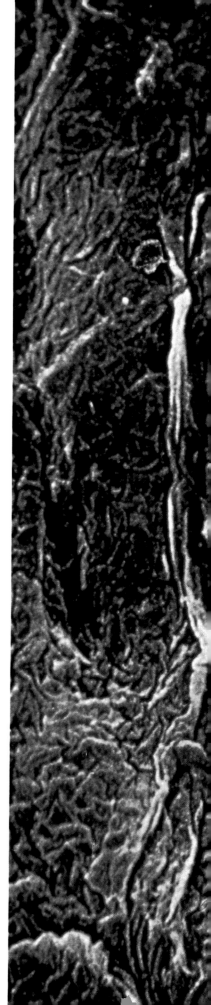

Marilyn J. Koering (below)

TOOTH. The beginning of a cavity looks like a crater (above). A tinted SEM picture shows a tiny spot of tooth decay. As bacteria bore into the enamel, the hard outer covering of a tooth, the enamel gradually dissolves and collapses.

HAIR. Magnified by the SEM, a single strand of hair looks like a tree trunk (right). The outer layer of the hair resembles torn strips of bark wound tightly around the trunk. Hair isn't alive. That's why a haircut doesn't hurt.

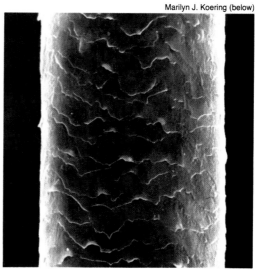

Lennart Nilsson from BEHOLD MAN; Little, Brown and Co., Boston

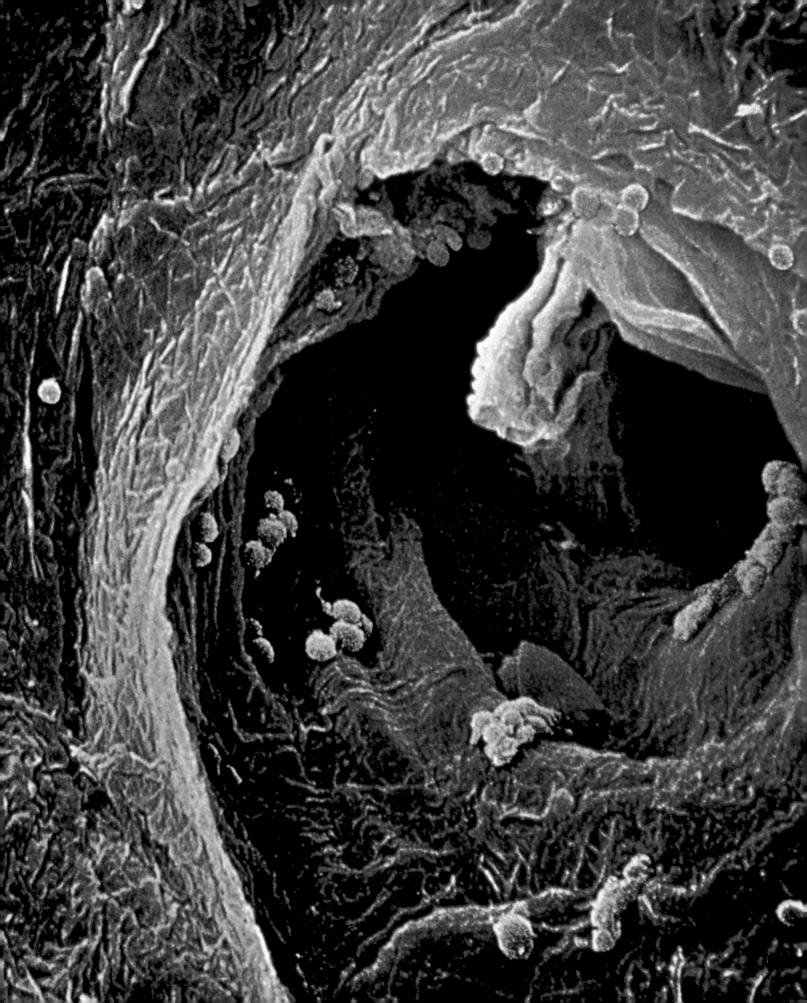

AROUND THE HOUSE

Magnifying tools give you surprising views of things you see every day.

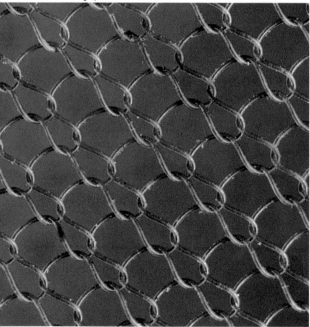

MAKING MUSIC, *the tip of a phonograph needle moves along a groove in a record (right). The groove was cut during a recording session. Each bump represents a sound. As the needle moves along, it translates each bump into an electrical impulse. The impulses go to an amplifier, which turns them into the sounds you hear. Color has been added to this* SEM *enlargement.*

Photographics/Earl A. Kubis

CHAIN-LINK FENCE? *No, it's the fabric of a nylon stocking (above). Here you can see how a single thread forms all of the links in the chain. When one link breaks, the connecting links start unraveling. Soon a tear runs up, or down, the stocking.*

Fred Ward/BLACK STAR

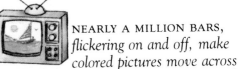
NEARLY A MILLION BARS, *flickering on and off, make colored pictures move across a lighted TV screen. The red, green, and blue bars (below) are made of materials that can glow. Each bar lights up when touched by a fast-moving beam of electrons keyed to that color.*

RCA

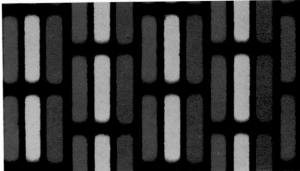

David Scharf from Peter Arnold, Inc.

SPAGHETTI SURFACE. *Matted fibers crisscross the surface of a sheet of writing paper (above). To the eye, the surface would look smooth. The fibers you see are wood. In papermaking, logs are chopped into bits, mixed with chemicals, and cooked into a soupy mass. The soup is sprayed onto a screen. Then it is rolled and dried, forming paper.*

25

2 BEYOND EARTH

*by Kathryn Allen Goldner
and Carole Garbuny Vogel*

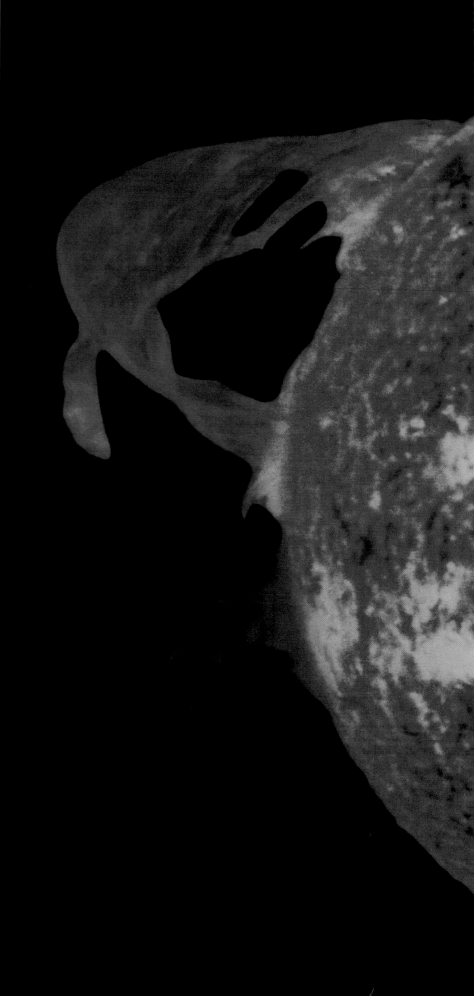

In ordinary photographs, the sun doesn't look very lively. It's just a small yellow disk in the sky. But photographed from an orbiting space station, the sun bursts with activity (right). Violent storms erupt on its surface. Great flares of glowing gases shoot far into space. Shock waves rush toward earth and the other planets.

You need special equipment to observe such solar activity. The sun is too far away for its surface details to be visible from earth. More important, it is too bright to observe safely. Looking at the sun directly would damage your eyes. Now, with the aid of telescopes, computers, and spacecraft, scientists can study the sun, which is the star closest to earth. They can also study more distant stars, and the planets. What do they see? You'll find out on the next pages.

HUGE LOOP *of superhot gases called a solar flare bursts from the sun's surface. Caused by magnetic disturbances on the sun, solar-flare activity peaks about every 11 years. Flares send shock waves into space. A shock wave can cause storms on earth and disrupt long-range communications. This flare was photographed from Skylab, an observatory that once orbited the earth.*

26   Naval Research Laboratory and NASA

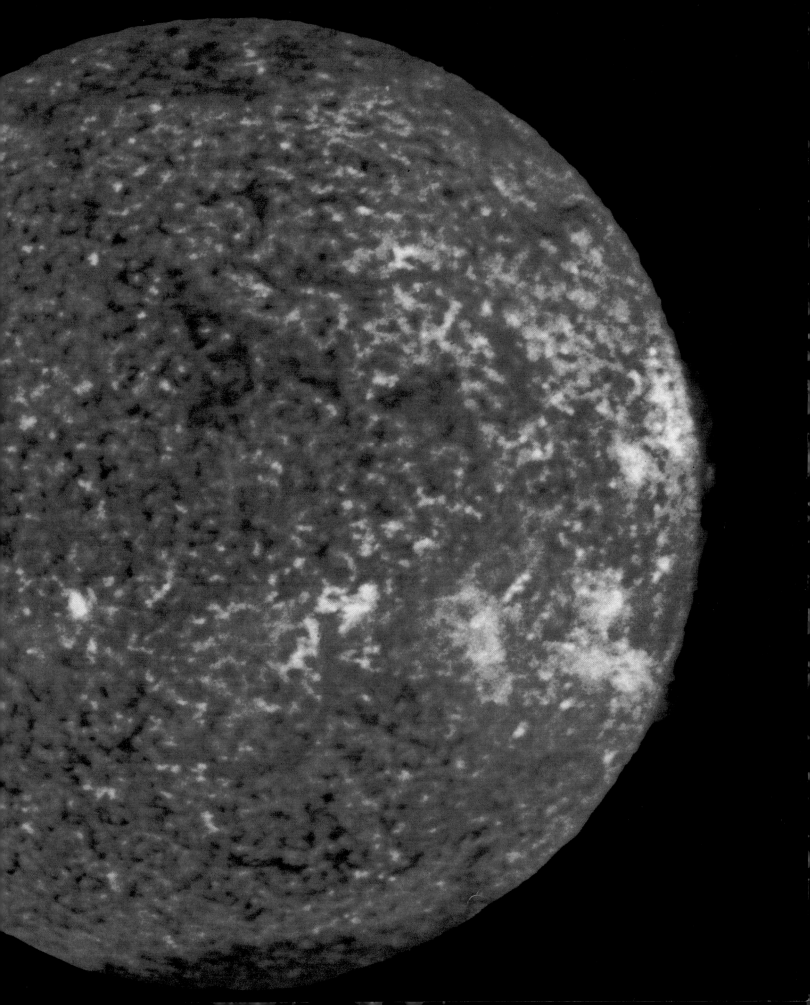

Distances in space are so great that astronomers use a special means of measuring them. Measuring the universe in miles would be like measuring the earth in inches. The mile is too small a unit to be used conveniently. Instead, astronomers speak of space distances in *light-years*. A light-year is the distance a particle of light travels through airless space in one year. To describe shorter distances, astronomers use *light-minutes* and *light-seconds*.

Light travels at 186,283 miles (299,793 km) a second. That works out to about 6 trillion miles (10 trillion km) a year. Look at the North Star some night. The light you see left the star 680 years ago—about 200 years before Columbus discovered America! If you wrote out the distance the light traveled in miles, you'd get this figure: 3,997,456,733,000,000. A number that large is hard to work with and harder to imagine. So astronomers say the North Star is 680 light-years from earth.

The sun is much closer. Sunlight travels to earth in about $8\frac{1}{2}$ minutes. So the sun is $8\frac{1}{2}$ light-minutes away.

Now, gear up your imagination. Come aboard a spacecraft for a trip that will take you toward the stars. You'll whiz into space at about 34,000 miles (54,700 km) an hour. Your destination is the Andromeda galaxy. Here's what you'll see as you travel farther and farther from earth:

Barbara Gibson (all)

ONE AND A QUARTER LIGHT-SECONDS. *After a trip of about seven hours, you land on the moon. You can see the earth at left. To its right are Venus, Mercury, and the sun.*

ONE LIGHT-YEAR. *You can still see the sun behind you. A small dot of bright light, it is the only visible body of the solar system. The planets are too faint to see. You notice a Voyager spacecraft that is moving through space on the same course you are following. You have both been traveling for 20,000 years.*

ONE HUNDRED LIGHT-YEARS. *The sun is no longer visible, but you see other stars. They glow in different colors, depending on their temperatures. The hottest ones glow blue. Yellow stars have medium temperatures. The coolest ones are red. Voyager is falling apart. Meteoroids—chunks of space rock—have battered it.*

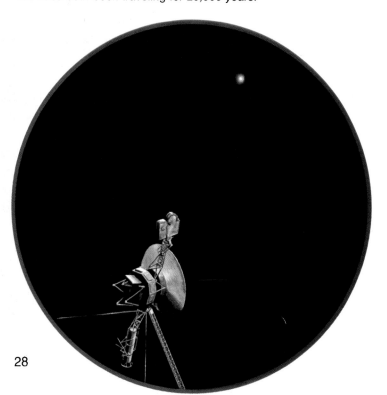

SIX LIGHT-MINUTES. *You reach Mars at the site of an earlier Viking landing. Above the planet's dusty red haze hang earth, Venus, Mercury, and the sun. The dot near earth is the moon.*

FIVE AND A HALF LIGHT-HOURS. *Your spacecraft lands on icy Pluto. Its moon, Charon, orbits nearby. The sun, still bright, appears small. Below it, you can see Venus, earth, Jupiter, and Saturn, which looks reddish.*

TEN THOUSAND LIGHT-YEARS. *An arm of our galaxy, the Milky Way, sparkles behind you. At lower left is a cluster, or grouping, of thousands of stars. At left you see two newly formed stars glowing next to a dark dust cloud. The lacy shape at right is what's left of an exploded star. A distant galaxy appears at lower right.*

ONE MILLION LIGHT-YEARS. *Halfway to Andromeda, you look back on the 70 billion stars that form the Milky Way. Star clusters, now so far away they look like single stars, form a halo around the galaxy. Voyager has been pounded to pieces and has disappeared. You have been traveling for 20 billion years!*

29

This is a recommended viewing device.

Exercise extreme caution in its use. **Never** directly view the sun. No liability is assumed or accepted by the producers and marketers of this device.

February 26, 1979

Lewistown

Montana

Eclipse Capitol Of The World!

WHAT'S THIS DOG DOING *with an eclipse viewer over its head? Its owner knows that it is dangerous to watch a solar eclipse without proper eye protection. Even during a total eclipse, the rays of the sun could burn unprotected eyes. You can safely look at an eclipse by holding a double thickness of exposed black-and-white film in front of your eyes. Use film that has been exposed to bright light for several seconds, then developed so it is totally black. If you are using a camera or a telescope, you must cover the lens with special filters.*
Jonathan T. Wright, Bruce Coleman Inc.

A vast cloud of glowing gases called the corona blankets the sun. Reaching hundreds of thousands of miles into space, the corona shines as brightly as a full moon. Yet you cannot normally see the corona. It is hidden in the even greater brilliance of the sun itself. It becomes visible only during a total eclipse of the sun.

A total eclipse is a rare event that lasts only a few minutes. But astronomers all over the world wait for that short period when the moon masks the sun. The sun's energy is constantly being filtered through the gases of the corona. Scientists study the corona to learn more about that energy and its effects on earth.

The next total eclipse of the sun visible from North America will not occur until the year 2017. Several partial eclipses will occur before then. During a partial eclipse, only part of the sun's surface is hidden. The corona remains invisible.

DURING A TOTAL ECLIPSE, *the moon hides the face of the sun (below). Only then does the corona become visible. Viewed from one place on earth, a total eclipse of the sun cannot last longer than $7\frac{1}{2}$ minutes. But in 1973, a group of scientists observed one for 74 minutes! In a speeding jet, they stayed in the moon's shadow as the shadow traced a narrow path across the earth. Studying the corona helps scientists understand weather patterns and predict disruptions of radio and TV signals.*

Jay M. Pasachoff

NASA/Solar Maximum Mission

SPECIAL CAMERA *aboard an earth-orbiting satellite made this picture of the sun's corona (left). The camera records ultraviolet, a form of light invisible to human eyes. To help scientists interpret the picture, a computer added color to the image. The picture shows how gases flow from the sun into the corona. A picture like this can be made only in space because gases in the earth's atmosphere filter out most ultraviolet rays. Read more about ultraviolet light in Chapter 5.*

Mount Wilson and Las Campañas Observatories,
Carnegie Institution of Washington

MOON MYTHS. *Before telescopes gave us a close look at the moon (right), people believed strange things about it. Some thought the moon was a shining toy that the sun god had stolen from a child to light up the night. Others thought that sleeping in moonlight or gazing at a full moon could make a person crazy. Telescopes and recent exploration show the moon to be a lifeless satellite made of dust-covered rock. It lacks both air and water.*

NASA/Apollo 17

FROM EARTH, *you could never see the moon the way you see it above. As the moon circles once around the earth, it also makes a complete turn on its axis, an imaginary line running through its core. As a result, the same side of the moon always faces the earth. The area hidden from earth is the lower half of the surface shown here. Astronauts made this picture as they circled behind the moon.*

MOON WATCHER. *Wendy Van der Woerd, 13, (right) peers at the moon through a telescope from the H. R. MacMillan Planetarium, in Vancouver, British Columbia, in Canada. The telescope magnifies the moon as much as 350 times. The instrument clearly shows craters, canyons, and ridges. Astronomers from the planetarium travel throughout British Columbia. Their programs give people a chance to look at the heavens through a telescope.*

32

David Barnes

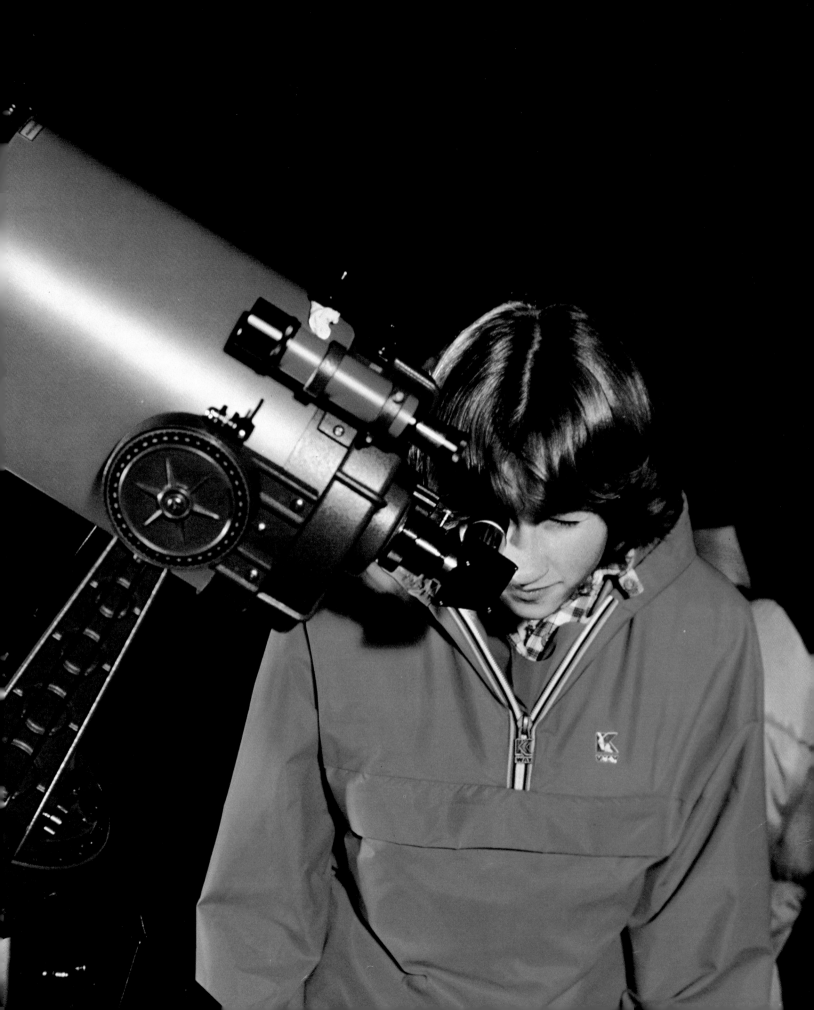

MILLIONS OF CRATERS dot the moon's surface. Meteoroids, chunks of space rock, punched them out. Many scientists believe meteoroids have been slamming into the moon for millions of years. Meteoroids fall toward earth, too. But as they hit earth's atmosphere, friction causes nearly all of them to burn up before they reach the ground. The moon has no atmosphere, so big and little meteoroids strike its surface.

Lunar Orbiter 3/NASA

When you gaze up at a full moon, you see only a bright disk marked by dark areas. The moon is about 240,000 miles (386,000 km) away—too far for you to see its surface in detail. If you look at the moon through a telescope, however, you can see many details of its face.

Deep, shadowy craters pepper the surface. At the rim of some craters you see high, craggy ridges. You discover that the dark areas are wide plains. Early observers called the dark areas *maria* (say MAR-ee-uh), the Latin word for "seas." They guessed that the areas contained water.

Scientists today know that the moon is dry. But mysteries remain. The far side of the moon has many more craters than the near side. The near side has more maria. Why do the two sides look different? Scientists don't yet know.

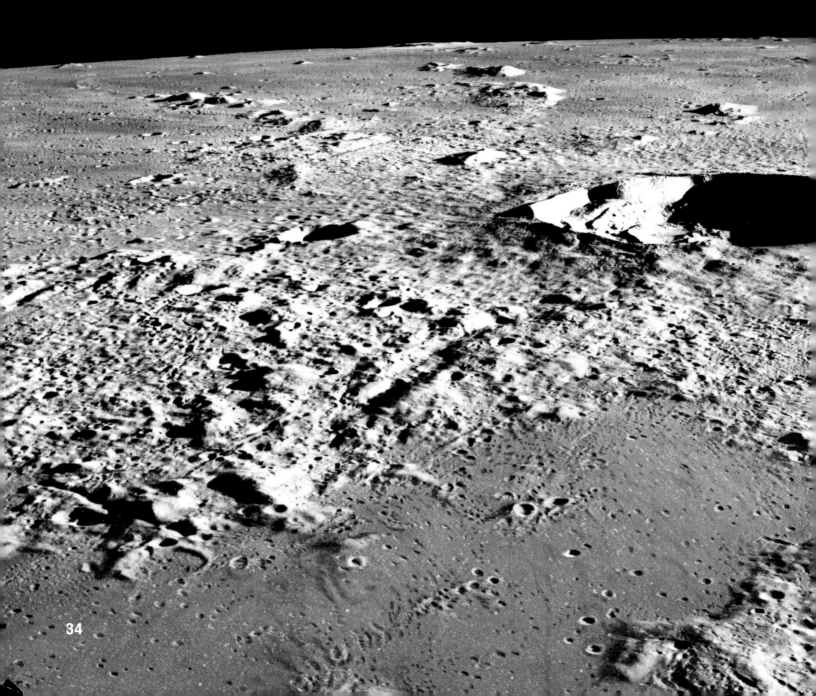

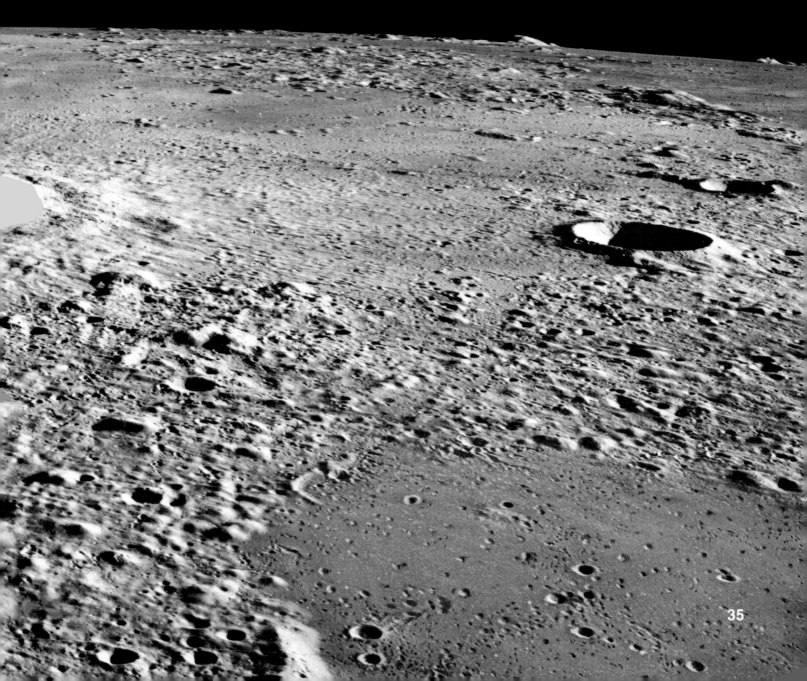

WHILE EXPLORING THE MOON *in 1973,* *Apollo astronauts left footprints on* *its dusty surface (left). There is no* *weather on the moon—no rain or* *wind to disturb the dust. As a* *result, the surface of the moon* *changes only very slowly. Hundreds* *of years from now, the footprints* *made by these astronauts may* *still be there, clearly visible to* *other moon explorers!*

NASA/Apollo 17/Eugene E. Cernan

35

SPEEDING PAST SATURN *in 1980, Voyager I took this picture (right) of the planet and its rings. Cameras aboard the craft recorded hundreds of rings that scientists hadn't known about. The rings are made up of frozen particles of chemicals. The planet itself is made up mostly of gases. But scientists don't yet know whether it is solid at the center. What lies inside is still a mystery.*

NASA/Voyager I

PORTRAIT OF THE PLANETS. *This painting shows the planets in their order from the sun, and compares their sizes (below). Closest to the sun is Mercury. It is followed by Venus, earth, and Mars, in that order. Next come the two giants, Jupiter and Saturn. Farther out are Uranus and Neptune.*

Pluto is the smallest planet. It is also the farthest from the sun. The sun itself is so big it could hold a million earths! This painting does not show how far the planets are from the sun or from one another. The space between them is really much greater than it appears to be here.

Barbara Gibson

NASA/Voyager I

LARGEST OF THE PLANETS, *Jupiter (above) could hold a thousand earths. The cloud-covered surface layer is made up of swirling gases. At lower left is the Great Red Spot, a violent storm that has lasted more than a hundred years. You can see Jupiter's moon Io at lower right. Voyager I discovered several erupting volcanoes on Io. The discovery surprised scientists. They had never detected active volcanoes beyond earth.*

"BEAUTIFUL, PEACEFUL LOOKING." *That's how one astronaut described earth (left) as he saw it from space. Featherlike clouds blanket much of the planet. Blue ocean, white polar ice, and brown land masses show through where there are no clouds. In this view, you can see the Arabian Peninsula, at upper right, and the African continent, at center.*

NASA/Apollo 17

37

Look at the sky on a dark, clear night. You'll see about two thousand stars. If you use a pair of seven-power binoculars, you'll see more than fifty thousand! There are actually trillions of stars in the sky. But most are thousands of light-years away. When they can be seen at all, it is only through a telescope. The sun, by contrast, is only $8\frac{1}{2}$ light-minutes away.

Stars are made of very hot gases that glow with brilliant light. The color of the light varies with a star's temperature. The hottest stars give off bluish white light. The coolest stars glow dull red. Yellow stars like the sun have medium temperatures.

Why do scientists study these distant objects? They believe there are many similarities in the lives of stars. By observing stars of varying ages, scientists hope to unlock clues to the past and the future of our sun, and of the planets that whirl around it.

SKY PICTURES. *Youngsters search the heavens (opposite) for constellations, groups of stars that seem to form pictures. There are trillions of stars in the sky. But even on a clear night, your eyes can detect only about two thousand of them.*

Ira Block (opposite)

Jay M. Pasachoff

EYE IN SPACE. *A scientist at a control center in Greenbelt, Maryland, views a television image recorded by a telescope aboard a satellite (above). Using the computer in front of him, he controls the orbiting telescope. The image shows a star named Capella. A computer has added color to the image.*

Bill Belknap/Photo Researchers, Inc./Rapho Division

PINWHEEL IN THE SKY. *Stars seem to trace paths through the northern sky (left). To make this picture, the photographer left a camera shutter open for about four hours. During that time, the earth turned on its axis, so the stars seemed to move. Their movements showed up on film as curved streaks of light. The North Star, almost directly above the North Pole, shows as the shortest, brightest streak.*

STREAKING PAST EARTH, *Comet West shoots tail-first away from the sun (below). The comet appeared in 1975–76. Its lopsided orbit around the sun takes it so far into space that it won't return for a million years! Comets approach the sun head-first. After rounding the sun, they retreat into space tail-first.*

SPIRAL-SHAPED GALAXY *named Andromeda stretches 120,000 light-years across space (right). It contains 200 billion stars. Our galaxy, the Milky Way, also has a spiral shape. Andromeda is one of the closest galaxies to the Milky Way. It is 2 million light-years away. The galaxy is greatly magnified here. To the eye, Andromeda appears as a small, faint blur.*

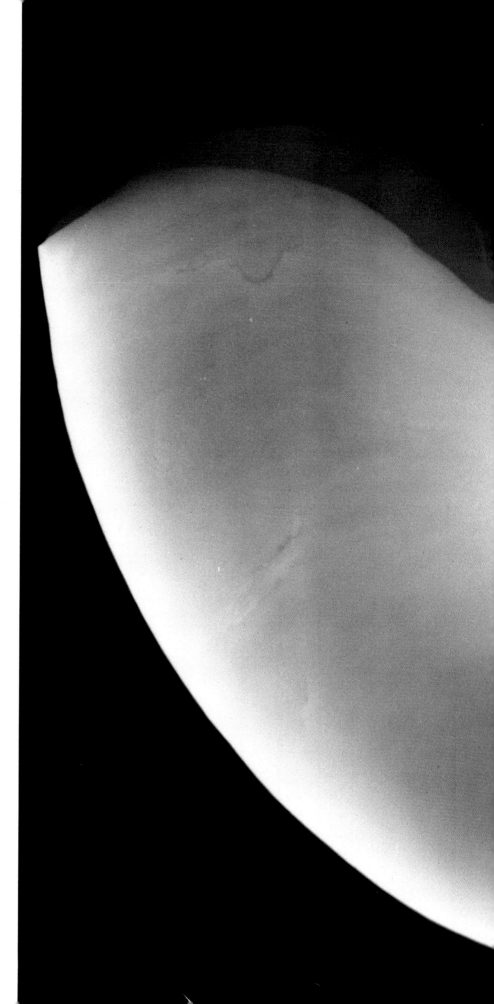

3 THE INSIDE STORY

by Catherine O'Neill

 colorfully wrapped box sits on the table next to a birthday cake. You can't tell what's inside the box, and you have to wait to find out. At a time like this, you may wish you had X-ray vision. Then you could see through the wrapping to the gift inside!

Of course, no one really has X-ray vision. But, by using special instruments, people can look through the surface of things to learn what's inside. Doctors look at the bones inside the body. Manufacturers check products for hidden imperfections. Scientists study the inner structure of seashells.

X rays are just one means of seeing

X-RAY PICTURE *shows the inside of a chambered nautilus shell. The nautilus, a soft-bodied sea creature, adds to its shell at the open end as it grows. Each time it moves forward into roomier quarters, it builds a wall behind its body. This picture is twice life-size. To see what a living nautilus looks like, turn the page.*

42   Robert P. Carr

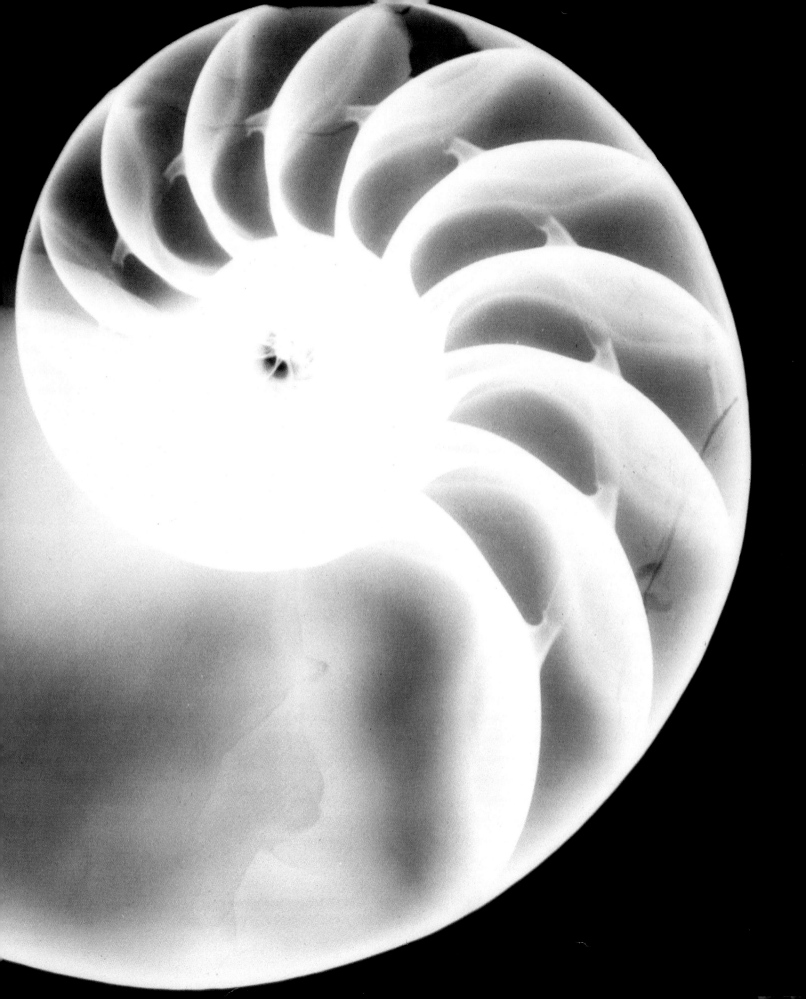

NAUTILUS PEERS *from the opening of its shell. The nautilus has as many as 90 small arms, or tentacles. Some grasp and hold the creatures it eats: crabs, fish, and lobsters. Others are used to smell things. The nautilus makes a substance in its body that hardens and forms its shell. Each time the creature moves toward the opening in its shell, it gradually pumps liquid out of the chamber behind it and fills that space with gas. The gas makes the nautilus light in the water, so it can move about easily. You can see the tube that carries the liquid and the gas in the X-ray picture on page 43.*

beyond the surface of things. Scientists have developed other tools as well. Light-carrying threads called optical fibers send out images from inside a human body or a machine. Sound waves are used to explore worlds we cannot penetrate with our vision.

A German scientist named Wilhelm Roentgen discovered X rays by accident in 1895. He was experimenting with an electron tube that was enclosed in a black box. He noticed that whenever he ran an electric current through the tube, a chemically coated screen nearby lit up. When he placed his hand between the box and the screen, the bones cast their shadow on the screen!

Roentgen called his discovery X rays. Scientists use an "X" to stand for something unknown, and Roentgen did not know exactly what X rays were. Later, scientists learned that they are a form of invisible light that can go through many objects.

Different kinds of light rays behave (Continued on page 48)

POCKET FULL OF WHAT? *You know that the person wearing the jeans has a comb and a pen in the back pocket (right). But you can't see what else might be hidden between the layers of denim. An X-ray picture tells the inside story (far right). The picture was made through a process called radiography. A technician put the jeans on a table, between film and an X-ray machine. Rays from the machine passed through the cloth, revealing the pocket's contents and making an image of them on the film. What did the X rays discover? A paper clip, a set of keys, two coins, and a pocket calculator. X-ray pictures are sometimes called radiographs.*

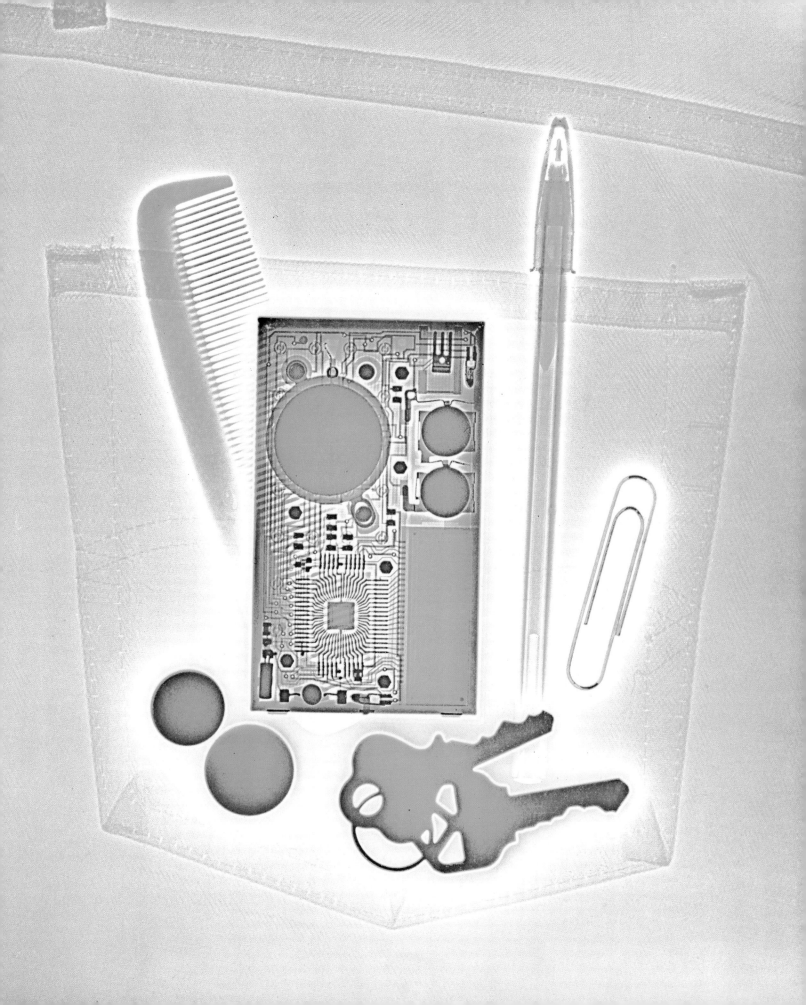

WHAT'S INSIDE A CB? *This radiograph (right) shows the transformers, wires, and other metal parts inside a citizens band radio. The small dots are dabs of solder (say SAHD-er), metal used to make electrical connections. Copper wire winds in a spiral inside the microphone cord.*

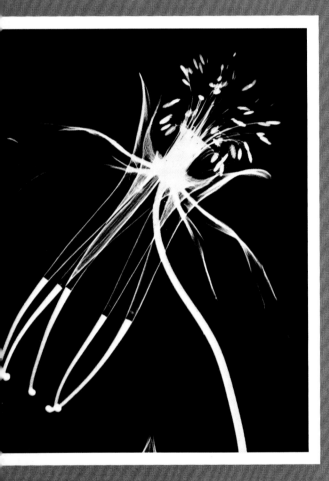

DANCING PHANTOMS? *A fireworks display? You might imagine you see either of those things in this radiograph of a wild flower called a columbine (above). The ghostlike objects are petals. Thicker parts of the plant show in solid white. Scientists use radiographs to study a plant's structure. Artists use the images for inspiration.*

46

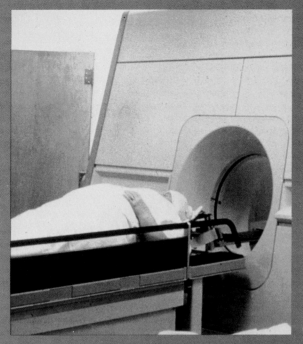

BRAIN SCAN. *A patient lies near an instrument called a computerized axial tomography (CAT) scanner (left). He's waiting to have radiographs made of his brain. They will show the brain in a series of images. Here's how it's done. An X-ray transmitter circles above the patient's head. It focuses on and scans a thin layer of the brain. It repeats the process until it has scanned the entire brain. With each scan, an image shows on a screen (below). Here, a computer has added color to show detail. Fast and painless, CAT-scanning enables a doctor to look at the brain slice by slice, much as you see a loaf of sliced bread.*

Dan McCoy from Rainbow (both)

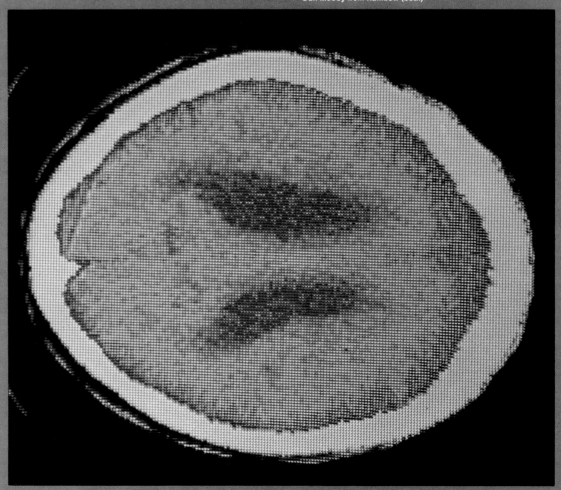

VOYAGE OF DISCOVERY. *A doctor has guided a tube through the twists and turns of a patient's digestive tract. The tube contains two bundles of optical fibers. Light travels through one bundle to the area to be examined. The second bundle sends back an image of the lighted area. Here, Dr. Jack Vennes and his assistants at the Veterans Administration Medical Center in Minneapolis, Minnesota, view the image on a TV screen. They are examining a canal that carries digestive juices to the small intestine.*

(*Continued from page 44*)   in different ways when they meet an object. Visible light rays often bounce off the surface of an object and into your eyes. When that happens, you see the surface, but not what's behind it. Instead of bouncing off an object, X rays pass through it. They can penetrate wood, cloth, metal, flesh, and many other materials.

X rays do not penetrate all materials equally well, however. For example, they pass through bone less easily than through flesh. Where X rays hit film, the film turns black. Since fewer rays penetrate bone, bone shows up on film lighter than flesh does.

Now scientists have found a way to focus X rays on the tissue behind bones. They have developed an instrument called the computerized axial tomography (CAT) scanner. It can pass X rays through the skull to make images of the brain in slices.

Doctors also explore the body through a viewing device that

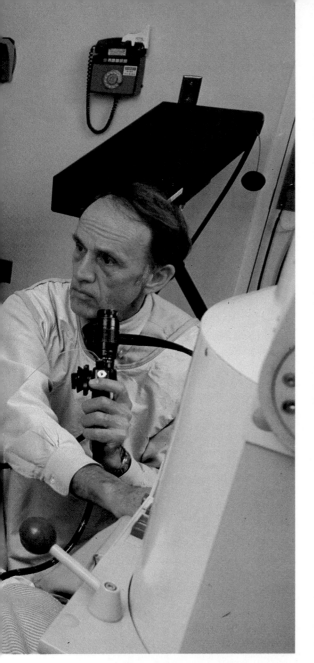

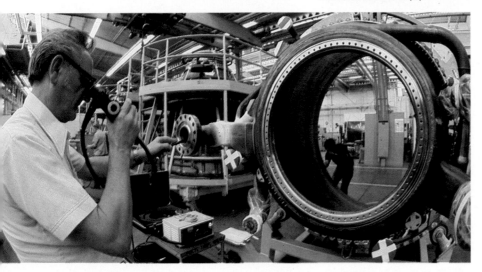

FOUNTAIN OF FIBERS. *Thinner than human hair, optical fibers (above) are made of flexible, clear glass. Light travels from one end of a fiber to the other, much as water passes through a hose.*

CHECKING FOR DIRT, *a technician at the Rockwell International Corporation in Canoga Park, California, inspects a nozzle for a space-shuttle engine (left). He looks into a viewer attached to a bundle of optical fibers. Inside the engine, a lens at the end of the bundle sends back the image he sees.*

49

uses optical fibers. The fibers are made of glass or plastic. Each fiber is as flexible as a human hair—and even thinner. Thousands of fibers are arranged in two narrow bundles. A doctor passes the bundles through the mouth, or another body opening, and into the area to be examined. One bundle sends light into the area. The second bundle carries back an image of the lighted area. Peering through an eyepiece, the doctor sees the image. With optical fibers, a doctor can examine internal organs without surgery.

Optical fibers are used in industry, too. Inspectors in automobile plants use them to examine the insides of engines. They can see whether the engines have been properly assembled.

Pictures of unseen things can also be made with sound waves. A device called sonar sends out high-pitched sound waves. The waves travel until they hit an object in their path. Then they bounce off the object, causing echoes. A computer catches the

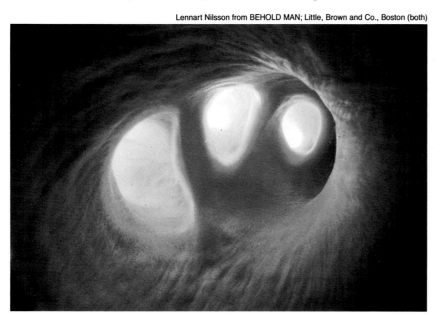

HIDDEN PASSAGES *inside the human body look like deep, winding caverns when lighted and magnified. The aorta (say ay-OR-tuh), largest of the blood vessels (above), receives blood from the heart. Arteries, the brightly lighted areas, carry blood from the aorta to all parts of the body. The middle ear (right) contains delicate bones. They vibrate with the eardrum—the thin skin at lower left. The bones send sound to the inner ear. From there, nerve endings carry it to the brain. Here you see the bone called the stirrup. It resembles the stirrup that supports a horseback rider's foot. Optical fibers provided the light for both of these pictures.*

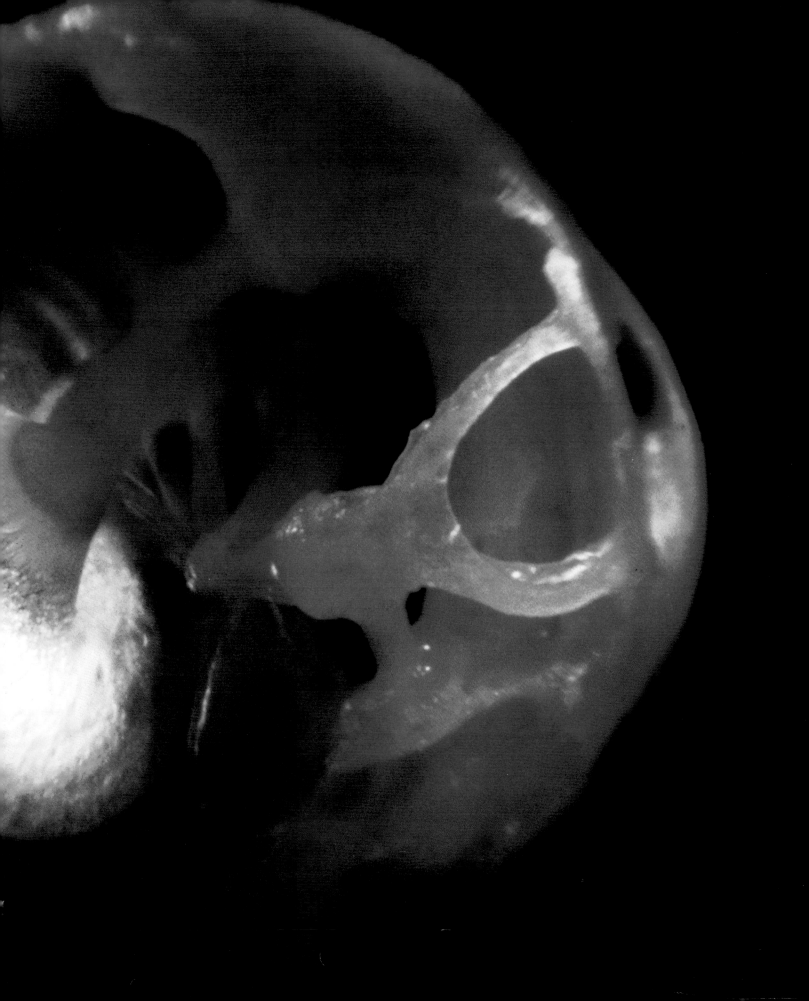

SUNKEN SHIP. *Dark image and white shadow (right) show the American warship* Hamilton *sitting at the bottom of Lake Ontario, in Canada. Sunk in a storm during the War of 1812, the* Hamilton *was lost for more than 150 years. Then, using sonar devices, scientists found it. Sonar sends out high-pitched sound waves and picks up returning echoes. The echoes are made when the waves bounce off an object. Here the echoes, changed electronically into an image, show the outline of the* Hamilton. *Experts say the ship is in good condition. They hope to raise the* Hamilton *and put it on display.*

The Canada Centre for Inland Waters

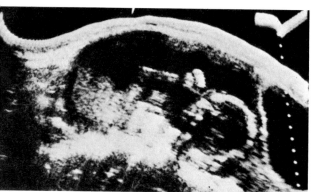

Edward R. Lipsit/George Washington University Medical Center

BABY PICTURE. *Sonogram shows an unborn baby 21 weeks old (above). Such pictures are made from reflected sound waves. Fast and safe, the technique helps doctors check a baby's development.*

SONAR HELPS *in the search for a long-lost painting by Leonardo da Vinci. Dr. John Asmus and his assistant, Maurizio Seracini (right), analyze echoes from a mural in Florence, Italy. From the echoes, they can tell if an older painting lies hidden beneath the one they see. This time, none did. Later, other researchers, using sonar and infrared devices, did detect one. Tests will tell if it is Leonardo's work.*

52

Gianni Tortoli

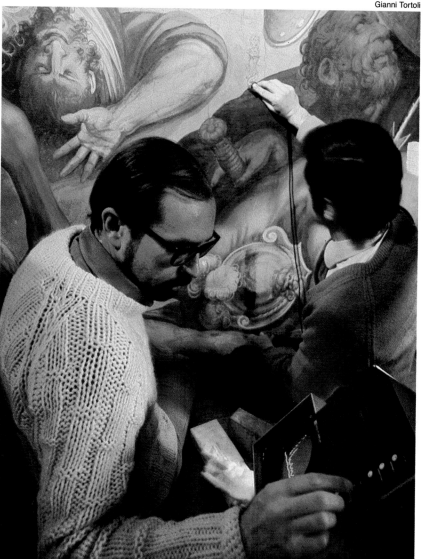

echoes. It changes them into signals that show up on a TV screen. Doctors use sonar to see if an unborn baby is developing normally. Art experts use it to help them find valuable old paintings that are hidden under layers of newer paint. Marine researchers use it to locate sunken ships.

Most of us, of course, don't have the chance to use special equipment to explore the insides of things. We have to rely on our eyes. But we can see what's inside the human body by examining plastic models on display at many museums.

One such display appears at the Museum of Science and Industry, in Chicago, Illinois. There, visitors view a model of a woman with transparent skin and flesh to learn about organs inside the body. Visitors also walk through a large model of the human heart. Check museums near your home. They may have similar models of hidden worlds for you to explore.

SUPERSIZE MODELS *give museum visitors a close look at how the body's internal organs work. A nephron (below), one of millions of filtering units in the kidneys, shows how the blood is cleansed of impurities. The body has 2 kidneys. They cleanse the body's entire blood supply 15 times an hour.*

ENTER   ENTER

PLASTER HEART *(left) tells the story of the body's most important muscle. Visitors walk inside this model. There, they see how the heart pumps blood to all parts of the body, carrying oxygen and food. These models are on display at the Museum of Science and Industry, in Chicago, Illinois. For a look at various organs in place inside a model of the human body, turn the page.*

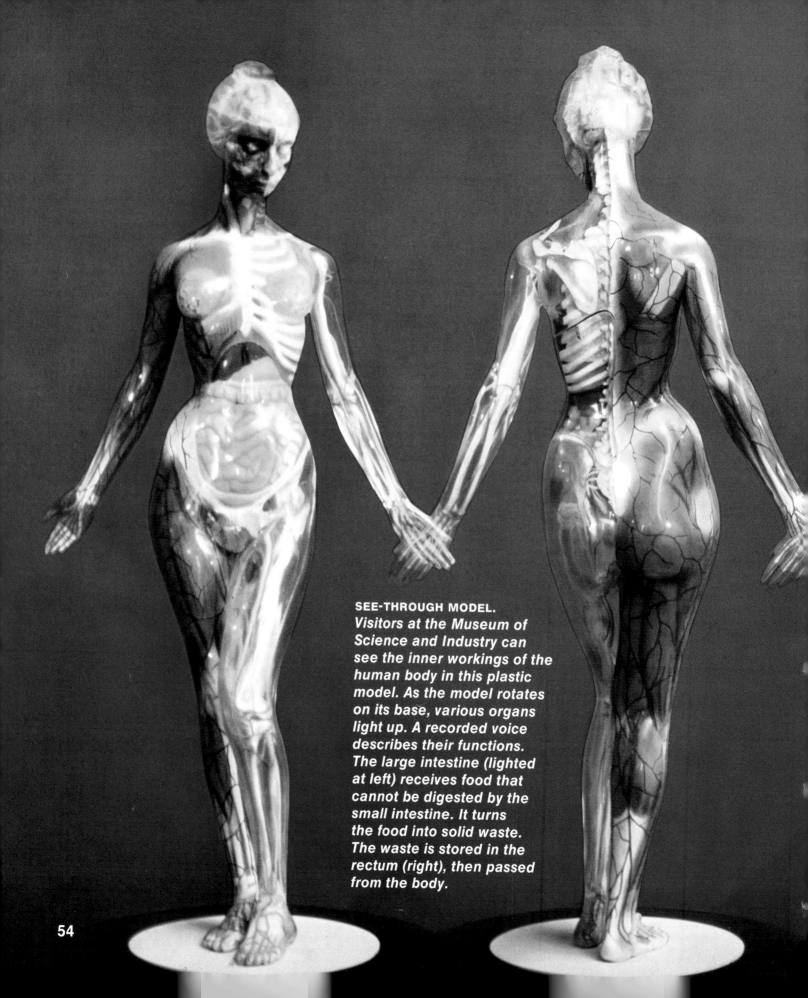

SEE-THROUGH MODEL.
Visitors at the Museum of Science and Industry can see the inner workings of the human body in this plastic model. As the model rotates on its base, various organs light up. A recorded voice describes their functions. The large intestine (lighted at left) receives food that cannot be digested by the small intestine. It turns the food into solid waste. The waste is stored in the rectum (right), then passed from the body.

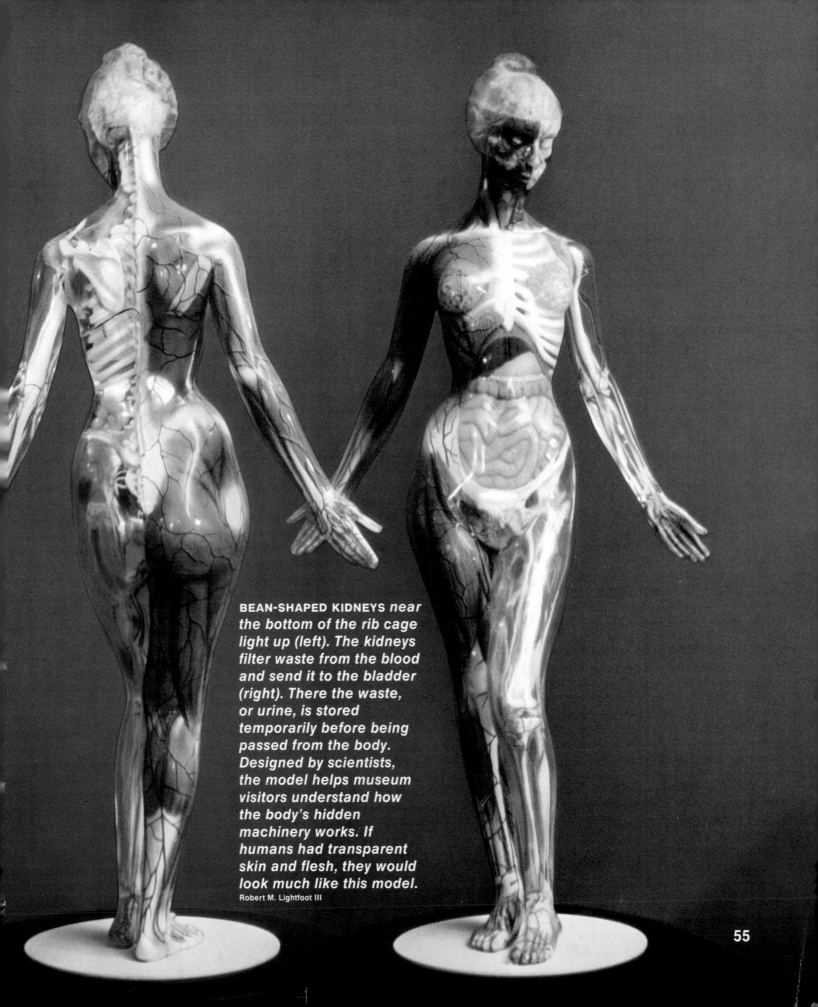

BEAN-SHAPED KIDNEYS *near the bottom of the rib cage light up (left). The kidneys filter waste from the blood and send it to the bladder (right). There the waste, or urine, is stored temporarily before being passed from the body. Designed by scientists, the model helps museum visitors understand how the body's hidden machinery works. If humans had transparent skin and flesh, they would look much like this model.*
Robert M. Lightfoot III

4 WIDER HORIZONS

by Eleanor Felder

ou are standing at the top of a hill with your dog. Both of you are looking at a sparkling blue lake in a meadow below. Does your dog see the scene as you do?

The answer is no. Different creatures see the world in different ways. Your dog sees little or no color. And its eyes do not focus as well as yours do. The gray images that it does see are mostly blurry.

But some creatures can see things you can't see. Can you spot a peanut on the ground from the top of a 120-story building? An eagle can! Can you see behind yourself without turning your head? A rabbit can! On the next pages, you'll find out more about the eyes of various creatures and the different ways they see.

HOW DOES A KITTEN KNOW *whether you're a friend? "I think it can tell by looking you in the eye," says Roger Hicks, 13, of Falls Church, Virginia. "That's what this cat did to me." The eyes of all vertebrates—animals with backbones—are similar to human eyes. But different animals see things in different ways. On page 62, you'll find out how a cat's vision differs from yours.*

56   N.G.S. Photographer Bates Littlehales

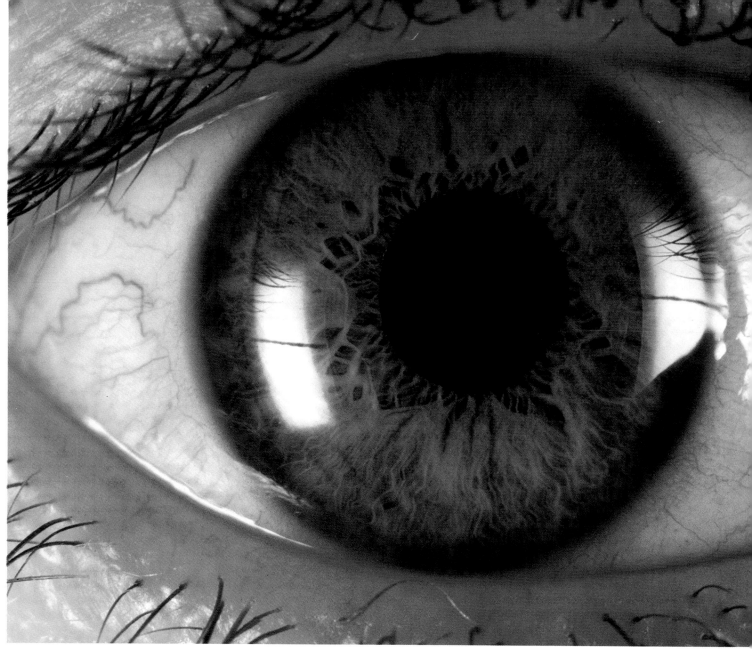

LOOK CLOSELY *at a human eye (above).* *The colored part is the iris. The dark* *opening in its center is the pupil. As the* *iris opens or closes, the pupil becomes* *larger or smaller. This movement* *regulates the amount of light that enters* *the eye. The sclera (say* SKLAIR-*uh), a* *white protective layer, covers most of the* *eye. Over the iris, the sclera becomes* *transparent and is called the cornea* *(say* KORN-ee-*uh). Unseen behind the* *eyelids are the tear glands. Tears lubricate* *the cornea. Blinking wipes it clean.*

Your eyes are remarkable organs. With them you can see a rainbow of colors. You can focus on a book in your hand, and an instant later on a speeding bike a block away. You can see at night, and you can judge distances. Your ability to see is the result of teamwork between the eyes and the brain.

An eye takes in light and focuses it. Light reflected from an object first passes through the protective cornea, where focusing begins, and through the clear fluid behind it. The light then travels through the pupil. The lens behind the pupil completes the focusing process. It sends the light through a clear, jellylike substance that fills the eyeball.

Finally, the light strikes the retina (say RET-nuh), at the back of the eyeball. There, millions of light-sensitive nerve cells change the light into electrical signals. The optic nerve sends these signals to the brain. The brain interprets the signals—and you see!

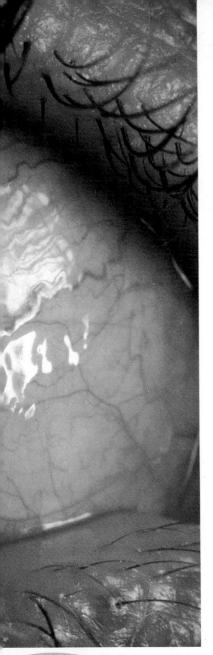

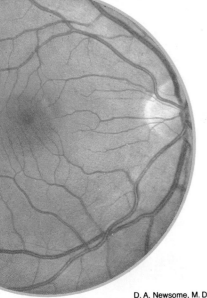

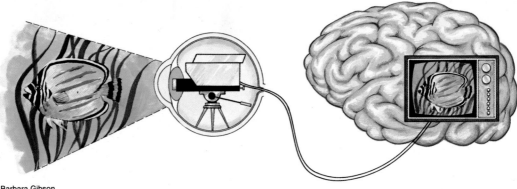

Barbara Gibson

THINK OF THE EYE *as a television camera (above). This will help you understand how vision works. The cornea and lens are like the lens of the camera. They focus light onto the retina. The retina works like the light-sensitive plate inside the camera. Both the retina and the plate change the* light into electrical signals. Then the optic nerve sends the signals to the brain in much the same way an electrical cable can carry signals to a television set. The brain translates the signals into the image you see, much as the television set changes signals into a picture.*

N. G. S. Photographer Bates Littlehales

NETWORK OF BLOOD VESSELS *lines the retina (left). The retina contains millions of nerve cells that are sensitive to light. They are most tightly packed at the fovea (say* FO-vee-uh), *the brownish area near the center. The fovea gives us our keenest vision. The yellow area on the right is one end of the optic nerve. To make this picture, the photographer used a lens that caused the retina to reflect a bright flash. Normally, the retina absorbs light.*

IT'S HARD TO OUTSTARE *an angelfish, as Jefferson Miers, 12, of Falls Church, Virginia, and Roger discover. "Fish don't blink," Jeff explains. "Most have no eyelids to blink with!" If you open your eyes underwater, the image you see will be blurry. But a fish sees clearly. In water, light rays bend differently from the way they bend in air. The lenses in the eyes of a fish have a round shape that brings underwater light rays into focus.*

59

EYES FRONT *is a basic rule in any race. But some horses are easily distracted. Three thoroughbreds (right) wear blinders as they gallop down a racetrack in Elmont, New York. Horses can see to the right and left, as well as straight ahead. Blinders prevent distraction by blocking a horse's side vision.*

EYES OF A HORSE *are among the largest in the animal kingdom (above). The size and the wide separation of the eyes give the horse a wide field of vision, or view. A horse can see two different images at one time, one for each eye. Or it can use both eyes to see one image, as people do.*

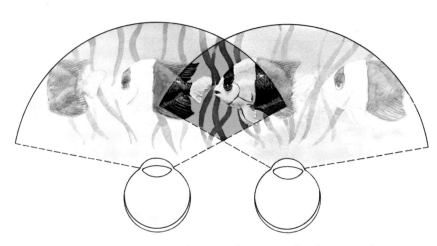

BINOCULAR VISION, *such as humans have, results from overlapping fields of vision (above). Each eye has its own field of vision. Where the two fields overlap, objects appear in sharp focus and have depth. Monocular vision occurs where the fields do not overlap. In these areas, objects are less sharply focused and appear flat.*

DIFFERENT KINDS OF EYES *have different fields of vision. Look at the drawing above. Blue represents binocular vision. Yellow represents monocular vision. Black shows blind areas. A human has a larger binocular field than a cat, but a cat sees in a wider angle behind itself. A rabbit sees in a full circle, except for two blind areas at very close range. Most of its vision is monocular, and so lacks depth.*

Barbara Gibson (both)

High in the sky, an eagle circles over a field. A rabbit in the field sits motionless near a hollow log. From a thousand feet (305 m) up, the eagle spots the rabbit. It swoops down. Although the eagle is approaching its prey from behind, the rabbit can glimpse the movement. It dashes inside the log—safe, for now.

The rabbit, the eagle, and other animals as well, have eyes that are especially suited to their survival. The rabbit's eyes are on opposite sides of its head. Often hunted as prey, rabbits need a wide field of vision. The placement of their eyes enables them to see in a full circle at one time. Their vision helps them spot danger from any direction. The rabbit sees two separate images, one with each eye. Such vision is called monocular (say muh-NOK-yuh-ler) vision. An animal with monocular vision has a broad view. But it can see little or no depth.

The eagle is a hunter. When an eagle   (Continued on page 64)

OUT ON A LIMB, gray tree frogs perch in the sun. The wide placement and the bulging shape of their eyes allow them to see in nearly a full circle around their bodies. When frogs sleep, they draw their eyes into their sockets. To close their eyes, frogs raise their lower lids. The lids, which help protect the eyes, are transparent. Through them, frogs can see moving objects, such as swooping birds. So even when they seem to be dozing, frogs can see danger coming.
R. Andrew Odum from Peter Arnold, Inc.

Stephen J. Krasemann from Peter Arnold, Inc.

SPREADING ITS WINGS, *a bald eagle gets ready for takeoff. It may have spotted a rabbit or other prey in the grass hundreds of yards away. Birds of prey, such as eagles, hawks, and falcons, have the keenest vision of all living creatures. They can clearly see distant objects that humans cannot pick out. Front-facing eyes give them excellent depth vision. With it, they can swoop down on prey and strike it with great accuracy.*

(Continued from page 62) spots prey, it swoops down, grabs the prey in its sharp claws, or talons, and flies off with it. To strike in exactly the right spot, the eagle must be able to judge distances accurately. Binocular vision—the combining of two images into one—gives it that ability. An eagle looks at an object directly in front of it, using both eyes. It forms two flat images that differ slightly because of the separation of the eyes. The eagle's brain combines the two images into one picture that has depth.

An eagle can see objects clearly at great distances. The retina in each eye contains about four times the number of light-sensitive cells found in a human retina. Look at a distant object through a pair of field glasses, first in focus, then slightly out of focus. The clear view will give you a rough idea of how much more keenly the eagle sees the world than humans do.

Vertebrates—animals with backbones (Continued on page 67)

ONE LANDSCAPE, THREE VIEWS. *An eagle, a human, and a dog see the same scene differently. To the eagle, the view is sharp, detailed, and colorful (below, left). The lens in an eagle's eye focuses light equally well onto large areas of the retina. So a large part of the field of vision is in sharp focus at all times. The retina itself has about four times as many light-sensitive cells as a human retina. That means an eagle's eye processes more information, allowing the bird to see in greater detail. A human eye can focus sharply, but only at one distance at a time (below, center). Only the light that strikes the extra-sensitive area called the fovea is sharply focused. The image on the rest of the retina is blurry. A dog's vision (below, right) is mostly blurry and colorless. It also lacks detail. But a dog's keen senses of hearing and smell make up for its poor eyesight.*

Robert E. Hynes

THOUSANDS OF TINY LENSES *packed tightly together make up the compound eyes of a horsefly (right). Each lens is aimed in a slightly different direction. Nerves carry a different image from each lens to the brain. The brain then puts the separate parts together into a complete picture. The bulging shape of the eyes enables the horsefly to see in nearly all directions at once.*

Dwight Kuhn (right)

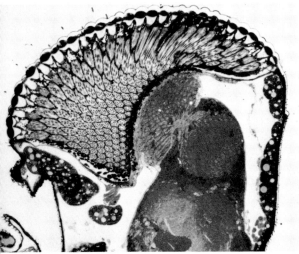

Courtesy of Dr. K. Hirosawa, Department of Fine Structure, University of Tokyo

INSIDE A FLY'S EYE. *A cross section shows the inner structure of a compound eye (above). The photographer got this view by slicing down through the eye of a dead fruit fly, somewhat as if he were slicing at a slight angle down a bundle of straws. The horn-shaped area across the top shows the lenses and the nerve structures that hold them. You can see a nerve bundle near the center of the picture. It has a butterfly shape. The bundle touches the brain below it. The brain is connected directly to the insect's muscles. This close link between eye and muscle helps the fly react quickly to danger—such as a flyswatter.*

(*Continued from page 64*) —have only one lens in each eye. The eyes of insects and of some other creatures have many tiny lenses, bundled together. Such eyes are called compound eyes. The number of lenses varies with the creature. It ranges from 7 lenses to 30,000! Each lens of a compound eye takes in a small part of the scene in front of it. Then the brain translates the many images into a recognizable pattern.

The eyes of most insects are on the sides of their heads. The compound eyes of crabs and lobsters are at the ends of movable stalks. Each stalk can aim an eye backward or forward, right or left. Both arrangements give the creatures a wide field of vision.

Compound eyes cannot change focus as the eyes of vertebrates can. As a result, the image they form is often not clear. But compound eyes are able to detect the slightest motion. That's important for catching moving meals!

Jeffrey L. Rotman

EYES ON STILTS. *The hermit crab (above) looks out on its underwater world through eyes at the tips of two slim stalks. The crab can move the stalks together or singly. One eye can peer backward while another looks forward. When the crab needs to see, it extends the stalks. At other times, it pulls its stalks back close to the sides of its head. Some other sea creatures, such as lobsters and shrimps, also have compound eyes at the ends of stalks.*

LOTS OF DOTS. *That's how a garden might look to a honeybee. Its compound eyes divide a scene into thousands of images. If you can't tell what these dots represent, slowly move the book farther and farther from your eyes. Flowers should appear. Find out in the next chapter how scientists think a bee sees colors.*
James L. Gould

69

5 MORE LIGHT ON THE SUBJECT

*by Kathryn Allen Goldner
and Carole Garbuny Vogel*

When you look at flowers in a garden, you see a variety of colors: red, pink, yellow, violet, and many other shades. A bee buzzing around the same flowers won't see them as you do. To the bee, red flowers look black. Other flowers take on ghostly colors and patterns that are invisible to human eyes. The bee probably will fly to some of these flowers, attracted by the patterns you can't see. You and the bee see color differently because your eyes are sensitive to different forms of light.

ULTRAVIOLET LIGHT *makes a yellow flower appear to be a different color. This kind of light is invisible to humans. But some insects, including bees, can see it. Ultraviolet rays are reflected by many flowers that are pollinated by bees. The rays help guide the bees to the nectar in the flowers. Turn the page to see this flower in ordinary light.*

70

Terry Domico/EARTH IMAGES

RAYS: VISIBLE AND INVISIBLE

TO HUMAN EYES, *the flower on pages 70-71 looks like this. It is a gazania, a member of the daisy family. Many daisies reflect ultraviolet light. Scientists think this makes them attractive to bees.*

Terry Domico/EARTH IMAGES

Light isn't something that stays in one place. It is constantly moving. It begins as bursts of energy given off by atoms, the invisible building blocks of everything in the world.

Tiny particles of energy stream out of the atoms and start traveling. They move in waves, like the waves in water. The space between their tops, or crests, is called the wavelength.

Wavelength determines what you see—and what you don't see. Your eyes are sensitive to only a small range of wavelengths. They don't react to others. But the eyes of some insects do. That's why a bee sees ultraviolet light, one of the things you can't see.

Although much of the light around you is invisible, you can see and feel some of its effects. Now modern instruments allow you to "see" some of this normally hidden world.

72   Barbara Gibson (right)

You probably have heard about the spectrum. Maybe you have seen a picture of it—a picture that looks like a rainbow. Perhaps you have seen how the rainbow forms. Your teacher may have aimed a beam of ordinary light at a triangular chunk of glass called a prism. The prism broke up the light into colors. These colors are commonly called the spectrum. But they are only part of it, the

GAMMA RAYS and X RAYS

Both X rays and gamma rays vibrate very fast and travel in short waves. Both can penetrate solid objects. Doctors usually use X rays to make pictures of hidden parts of the body.

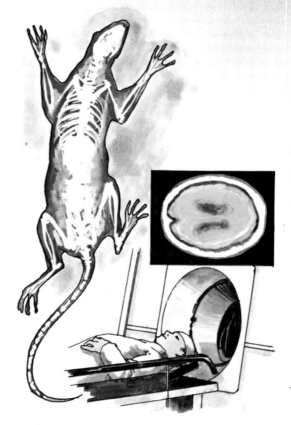

ULTRAVIOLET

Ultraviolet rays vibrate more slowly than X rays do, but not slowly enough for the human eye to see. These rays can be used to detect hidden paintings. They also cause sunburn.

visible part. All known light rays—visible and invisible—form a larger, electromagnetic spectrum. The name comes from the fact that every particle of light, or radiant energy, has an electric field and a magnetic field around it. The particles travel in waves of different lengths. Some radio waves are hundreds of meters long. Gamma waves are only a tiny fraction of a millimeter long.

Radar diagram adapted from The World Book Encyclopedia, Copyright © 1980, World Book–Childcraft International, Inc.

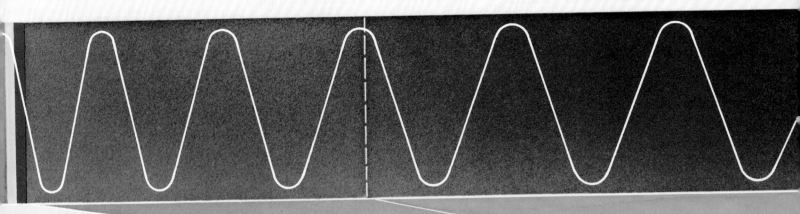

VISIBLE LIGHT

Ordinary light, called white light, contains the colors the human eye can see. When white light passes through a prism, the prism separates it into violet, blue, green, yellow, and red.

INFRARED

Invisible infrared rays can often be felt as heat. Infrared lamps keep food warm. Special instruments measure even infrared you can't feel. They make maps, like this one of a house leaking heat.

RADIO WAVES

The shortest radio waves are microwaves. They can cook food. Used in radar, they help police catch speeders. Slightly longer waves called shortwaves are used in long-range communication.

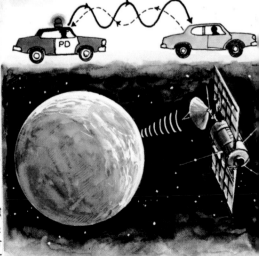

SPOOKY ULTRAVIOLET

ULTRAVIOLET RAYS *give minerals in a display case a ghostly glow (below). "These rocks are fantastic when they are lighted up," says Dana Kirkland, left. She and Eavonka Pollard, both 14, went with their eighth-grade class to the New Mexico Bureau of Mines Mineral Museum, in Socorro, where they studied minerals.*

Two white boxes sit side by side under a special lamp. They look exactly alike. Then the lamp is switched on. Suddenly, the boxes no longer seem identical. One looks yellow. The other looks white. How can that be? Each box is covered with a different kind of paint. Certain kinds of coloring matter make some white paint look yellow under ultraviolet light.

Since human eyes cannot see ultraviolet rays, their effects seem almost magical. As invisible ultraviolet rays strike nearby objects, the rays may make some of the objects produce visible light in colors different from those you normally see. Usually the objects give off a spooky glow that is blue, reddish, or greenish yellow. If an object glows only while ultraviolet light shines on it, we call it fluorescent (say floor-ES-uhnt). If it keeps glowing after the light is turned off, we call it phosphorescent (say fahs-fuh-RES-uhnt).

Ultraviolet rays have many uses. Art dealers use them to detect

N.G.S. Photographer Otis Imboden (all)

CHECKING ROCK SAMPLES, *Robert North, the museum director, shows students how to use a scintillation counter (right). It measures radioactivity. Certain minerals give off small, harmless amounts of radiation. The students are, from left, Valerie Martin, Shane Ohline, Ildiko Oravecz, and Sandra Sandman.*

ON A NEARBY PEAK, *other students find volcanic rocks about ten million years old (far right). Examining them are, from left, Karen Campbell, Keli Etscorn, and Joyce Vallejos. The students took the rocks back to the museum and studied them under an ultraviolet lamp. Some glowed, showing that they contained fluorescent or phosphorescent minerals.*

74

UNDER NORMAL LIGHT, *sometimes
called white light, a museum display looks
pale (below). Salt crystals coat the small
branches of a bush. This bush was once
covered by salty water. When the water
evaporated, the salt was left behind and
formed crystals. The rock in front of the
branches contains three minerals: calcite,
willemite, and franklinite.*

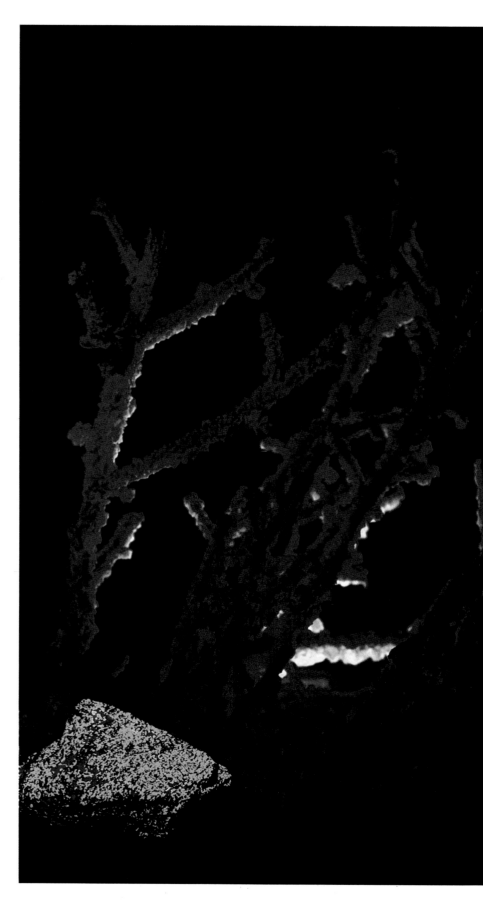

N.G.S. Photographer Otis Imboden (both)

PRESTO! CHANGE-O! *Ultraviolet rays
produce new colors (right). The salt and
calcite glow red. The willemite turns
green. The franklinite stays dark because
it does not fluoresce. Some minerals glow
with one color under an ultraviolet lamp
and take on a different color after the
lamp is turned off. Here, the lamp is on.
You can see it behind the lower branches.*

76

imitations of old paintings and to find old paintings hidden under newer ones. The colors that show up under ultraviolet lighting tell an expert the age of the paint. Modern paints glow with colors different from those of old paints. If a painting that is supposed to be old glows with a certain color, the dealer knows it is a fake. Ultraviolet also helps detectives spot forged signatures. The rays often make pen strokes of an erased or changed signature glow.

Ultraviolet rays are present in sunlight. They help your body make vitamin D, which builds strong bones and teeth. But too much ultraviolet is harmful. The rays can damage body cells. Doctors sometimes use small amounts of ultraviolet to treat skin diseases and to kill harmful bacteria. When you get a suntan, your skin is reacting to ultraviolet rays by building a dark screen to keep the rays out. If you don't give your body time to build such a screen, you will feel the result: a painful sunburn.

NOW YOU SEE IT . . . Ultraviolet makes a scorpion glow like a ghost. Its surroundings look dark. When white light shines on a scorpion, you see something quite different (bottom picture).

N.G.S. Photographer Bruce Dale (both)

. . . NOW YOU DON'T. Under visible light, the scorpion almost disappears (left). Its coloring blends with the earth and plants around it. Scorpions are found in many areas. This kind lives in very dry parts of the southwestern United States. Like most desert dwellers, scorpions escape daytime heat by hiding in cracks and holes. At night, they come out to hunt smaller creatures while darkness hides them. The scorpion would never look as it does above without special lighting. The chemicals that make its shell hard are fluorescent.

77

AMAZING INFRARED

HEAT MAP *shows how temperatures vary within a flame and in the air around it (below). The white area is hottest. Yellow and red are next. Pink and violet are cooler. To make heat maps, instruments measure infrared rays. They combine the measurements into pictures. Colors are added by computer.*

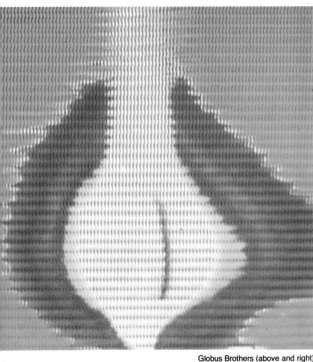

Globus Brothers (above and right)

ON THE SURFACE, *normal body temperature is really many temperatures, as the heat map (right) shows. Can you read it? The color code, from warmest to coolest, is white, yellow, red, violet, light blue, green, and dark blue. Doctors first studied body temperatures 2,400 years ago. They covered a patient's body with wet clay. The clay dried fastest on the warmest areas. An area that is unusually cool or unusually warm may indicate a problem inside the body.*

78

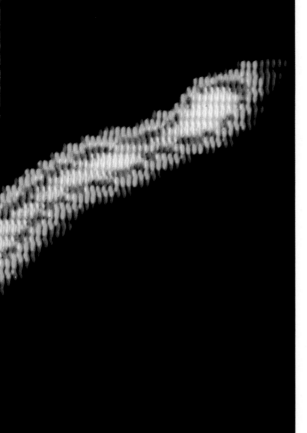

What does your body have in common with a rock, a flame, and a tree? All of these things give off infrared rays. Infrared was the first part of the invisible spectrum to be discovered.

In 1800, a British astronomer named William Herschel was using a thermometer to test the various colors found in sunlight. When the thermometer was placed beyond the deep-red end of the spectrum, it registered a higher temperature than visible light registered. He concluded that invisible rays were being produced. He named them infrared, which means "below red."

Every object on earth gives off infrared rays. Like visible light rays, they travel in waves of different lengths. Some waves produce a warmth you can feel. Most do not. You cannot feel the heat waves from a tree or a rock. But special instruments and certain kinds of film can measure and record this heat so that you can see it with your eyes.

National Environmental Satellite Service/NOAA

WATER TEMPERATURES *become visible in a photograph made from a satellite (above). Colors were added to show the differences. The warm Gulf Stream is a purple, red, orange, and yellow mass. Cooler waters around it are blue. Part of the United States forms a dark mass. Clouds show as white fluff. Heat maps have many uses. They help scientists predict the weather. They show places where buildings are leaking heat. Higher-than-normal temperatures give early warnings of forest fires and coming volcanic eruptions.*

DETECTIVE WORK. *You are searching for two dangerous escaped criminals. Could they be hiding in a nearby woods (right)? You can't quite tell. But you want to find out before going any closer. A thermograph can tell you. This sensitive instrument measures infrared waves. In just 30 seconds, it makes 10,000 measurements and turns them into a single color picture (far right). Two men hidden in the shadows now show up clearly. Their faces and uncovered arms give off the most heat (red). Their clothing gives off less heat (yellow). Trees and underbrush show as cooler colors (tan, green, blue, and black). The red-and-yellow strip in front of them is a paved road, warmed by sunlight. At night, the thermograph becomes an even better detective. As plants and the ground cool off in the cooler night air, warm bodies show up more clearly.*

Light in Darkness

LASER BEAMS *from a special TV camera scan the face of a woman. Although she is in a dark room, a bright picture shows up on the screen to her left. A laser is an instrument that produces very strong light. The light is called a laser beam. It moves in a straight line and can travel a long distance without growing much weaker. This is something most light cannot do. Methods of viewing objects in very weak lighting are known as image intensification. Scientists now have several ways of intensifying images. With image intensifiers in the form of special goggles, people who have certain eye problems, such as night blindness, can see things they couldn't see with their eyes alone.*

Howard Sochurek

POLARIZED PATTERNS

WEAK SPOTS *in a plastic drinking cup become visible under polarized light (below). This kind of light is produced by filters that block some light waves and let others through. Darker colors show where the cup is most likely to crack.*

Arnold G. Wilbur

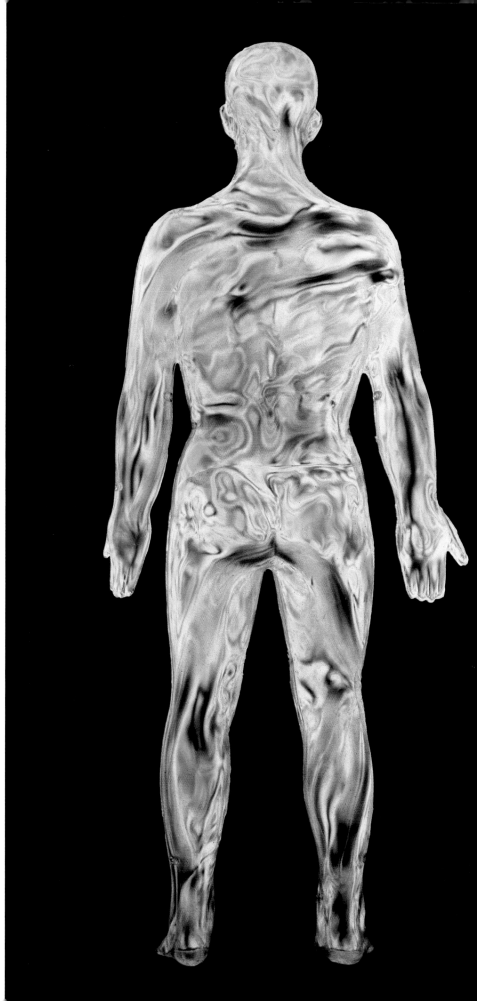

MODEL MAN *shows stress areas in a piece of plastic (right). Pictures like this one, made with polarized light, tell manufacturers where to look for weak spots in their products. Engineers often make small plastic models of buildings and photograph them using polarized light. Areas of stress and strain show up in the models. Builders can then strengthen these areas to make up for the weaknesses.*

Michael Freeman, Bruce Coleman Ltd.

Did you ever shop for a pair of sunglasses and notice tags on some that said "polarized" or "polarizing"? Such glasses do more than ordinary sunglasses. They cut down on glare by blocking light waves that are vibrating horizontally. Reflected glare is mostly made up of light waves that vibrate this way. Skiers and other athletes often wear polarized glasses to reduce eye strain.

A polarizing filter is a piece of glass, plastic, or some other transparent material that allows only certain light waves to pass through it. Waves vibrating in one direction pass through (right). Waves vibrating in other directions are blocked.

Scientists often add polarizing filters to their microscopes. When such filters are combined in special ways they make colors appear in normally colorless objects, such as salt and clear plastic. The added color shows the structure of the objects and helps people learn more about them.

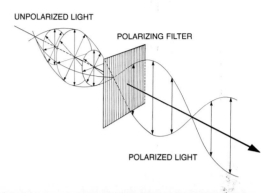

UNPOLARIZED LIGHT

POLARIZING FILTER

POLARIZED LIGHT

Manfred Kage from Peter Arnold, Inc. (both)

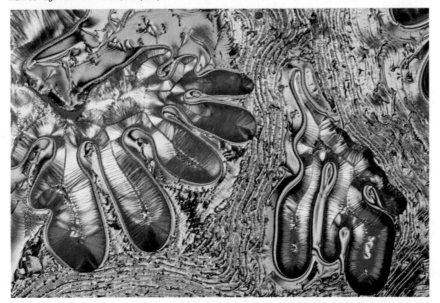

VITAMIN ART—B6. *You've seen vitamin pills many times, but have you ever seen vitamins as they appear on this page? A crystal of vitamin B6 (above) looks like a piece of modern art. It has been magnified by a microscope and colored by polarized light. This vitamin helps keep your skin and blood healthy.*

VITAMIN ART—C. *Polarized light makes colors swirl over the surface of a vitamin C crystal (left). This vitamin helps keep your bones and teeth strong. It is important in maintaining healthy gums and in healing cuts. This vitamin is also called ascorbic (say uh-SKOR-bik) acid.*

FROSTY "FLOWER." *During a storm, snowflakes look white, and all of them look alike. But if you could see one flake under a microscope with polarized light, you might see a flower shape (below). No other snowflake will look exactly like this one. To make this picture, the photographer coated a snowflake with plastic to form a mold. Then he photographed the mold after the flake had melted.*

Roger J. Cheng/Atmospheric Sciences Research Center, S.U.N.Y.A.

POLARIZED LIGHT *and magnification show the structure of a diamond (right). Diamond crystals form in many shapes. A common one looks like twin pyramids placed base to base. All the flat surfaces in this crystal are triangular. A diamond is the hardest natural substance. Some diamonds are used to cut glass and other hard materials.*

Fred Ward/BLACK STAR

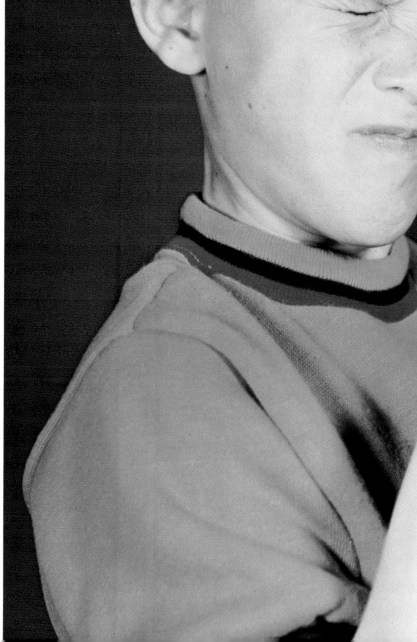

6 CATCH ALL THE ACTION

by Catherine O'Neill

You're having a picnic in the park. Birds fly overhead. Insects buzz near the picnic table. Some of your friends play ball, while others explore the woods for rocks, leaves, and flowers. You look around carefully and see a lot of things. But your eyes can't absorb all of the action that is taking place nearby.

If you brought along special cameras and lights to photograph your picnic, the pictures might show what your eyes miss. Cameras reveal things that human eyes can't see.

Some things happen too fast for our eyes to follow. But a camera can catch the action. Some things grow, move,

BANG! *A balloon bursts. Even if this boy had his eyes open, he wouldn't be able to see what the picture shows. A bright flash of light and fast film stopped the action for the photographer just as the skin of the balloon pulled back from the spot where the pin pricked it. Reflected light from the flash makes bright spots on the inside surface of the balloon.*

86   C. E. Miller/Strobe Lab, M.I.T.

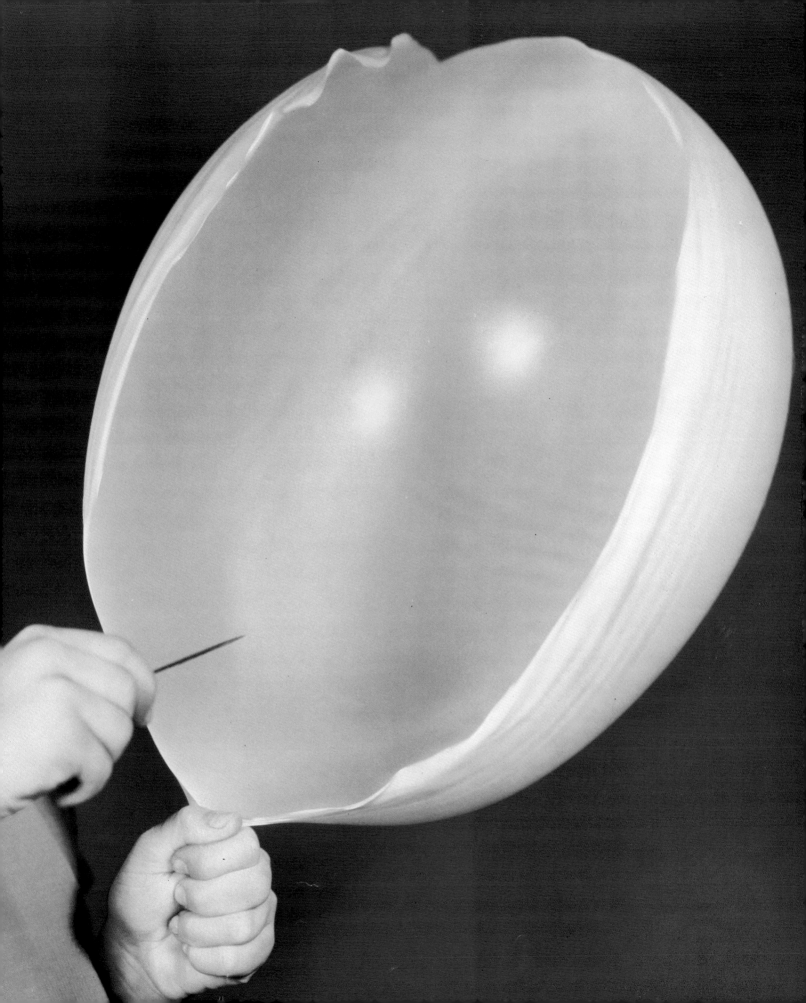

or change too slowly for us to notice. A camera can capture these changes on film. To our eyes, a ball whizzing through the air becomes a blur. We can't see the individual feathers on a bird's wing as the bird flies by. We miss the small, quick movements of a friend's muscles as the friend hits a home run during a ball game. We can't watch the grass grow, or a flower unfold its petals. Cameras can show us all of these things—and much more.

To make pictures like the ones in this chapter, photographers use special equipment and methods. High-speed photography uses short flashes of light to "freeze" fast-moving things. Time-lapse photography makes many images over a period of time that may range from minutes to months.

The word "photography" comes from two Greek words meaning "light" and "drawing." Light changes the sensitive surface of film in a camera, forming an image on the film.

Most cameras have a button that you push to take a picture. The button works a shutter inside. As the shutter opens, light reflected from objects outside enters the camera. The light strikes

CAUGHT . . . ON THE WING. *Have you ever tried to swat a fly as it buzzed around the house? If you have, you know it can get away from you very quickly. Its wings move so fast that all you see is a blur. A stop-action photograph (above) seems to freeze a housefly in the air. Now you can see how it looks in flight.*

BURSTING THROUGH AN APPLE, *a bullet moves so rapidly the eye can't see it. But a flash of light shows the bullet and the exploding fruit (right). The bullet whizzed past the camera at 1,800 miles (2,897 km) an hour. The light flashed for only a third of a microsecond. (A microsecond is one millionth of a second.)*

88

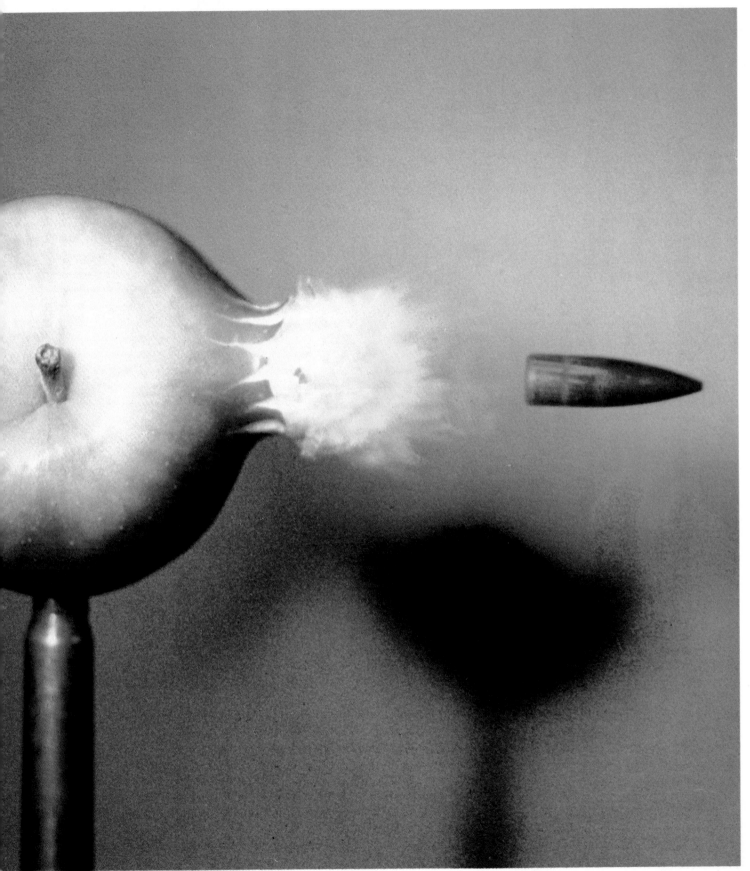

GIANT ACROBAT. *Dusky dolphin flings itself out of the water, then nose-dives. This action, called breaching, happens very quickly. A stop-action photograph, made near New Zealand, caught this dolphin in the middle of its leap. By studying such photographs, scientists can learn more about how large sea mammals use their bodies during their playful out-of-the-water acrobatics.*

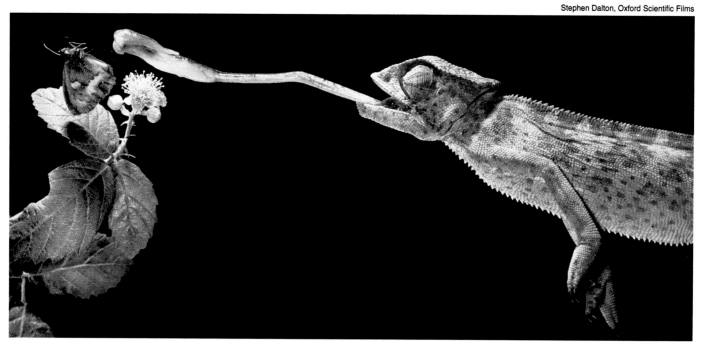

GOTCHA! *Stop-action photograph shows a chameleon at the instant it strikes a butterfly. This lizard has a tongue longer than its body. Most of the time, the tongue stays compressed inside the mouth. To catch prey, the chameleon shoots its tongue out in a fraction of a second. It strikes an insect and usually grabs the victim with its sticky tongue. Then it pulls back its tongue. If the insect is attached, the chameleon has a meal.*

the film and forms an image. Then the shutter closes. The finished picture will show whatever was happening at the moment the light came in and touched, or exposed, the film.

If you have ever photographed something that moved quickly as you pushed the button, you probably discovered that the picture came out blurry. Your camera shutter did not open and close quickly enough to produce a clear picture of a fast-moving object.

For years, photographers struggled with the problem of stopping action on film. Gradually, they developed equipment that helped them do it. On many automatic cameras, the shutter stays open for a hundredth of a second. But some cameras have shutters that open and close in *a thousandth* of a second! To capture an image

SWOOPING BARN SWALLOW drinks as it skims over the surface of a stream in the English countryside. For years, scientists have argued about how birds use their wings and bodies during flight. Photographs like this one have helped answer some of their questions. By studying birds, people have learned a lot about the mechanics of moving through the air.

Stephen Dalton, Bruce Coleman Ltd.

Brian Payne

PHOTO FINISH! *Amy Sunde, 17, of Colorado Springs, Colorado, sprints along a track at the Olympic Training Center there. As she dashes by, Rocco Petitto films her with a high-speed movie camera. Athletes and their coaches study such film with the aid of computers. By analyzing their own movements, athletes can learn what to do to improve their performances.*

on film in a thousandth of a second, you need supersensitive film, called fast film, an extra-bright light, or a combination of these things. An electronic flash produces light that is much brighter than daylight. The flash goes off after the camera's shutter opens. Whatever takes place during the split second of the flash appears on the film. Fast camera shutters, fast film, and bright lights allow photographers to make pictures of such things as lizards catching insects with their tongues.

Fifty years ago Dr. Harold E. Edgerton, a scientist at the Massachusetts Institute of Technology, in Cambridge, Massachusetts, wanted to make pictures of some machines he was studying. But the parts of the machines moved too quickly for any existing camera to get a clear picture of them. Dr. Edgerton invented a high-speed flashing light called a stroboscope. You may have seen strobe lights in a disco or in a (Continued on page 97)

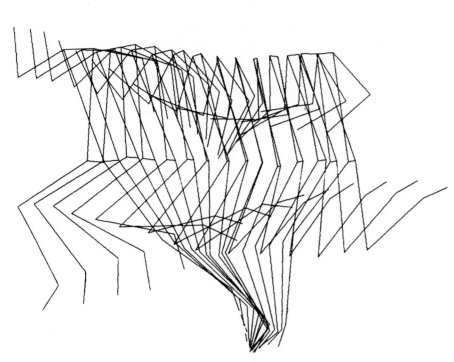

COMPUTER'S EYE VIEW *of a runner (left) shows how the runner's feet, legs, body, and arms work together. A computer translates high-speed movie film into drawings like this one. The computer's memory contains information on an ideal runner's style. Comparing the real athlete with the ideal one helps a coach spot small mistakes that the eyes miss.*

UP, UP, AND AWAY! *A computer printout (below) shows a javelin thrower in action. This sport requires split-second timing. An athlete must release the javelin at exactly the right moment, when it is in the best position to travel a long distance. A javelin thrower makes one fast motion. The computer drawing separates the motion into many small steps. Each step can be studied—and improved, if necessary. The use of computers to analyze film is part of a new science called biomechanical analysis.*

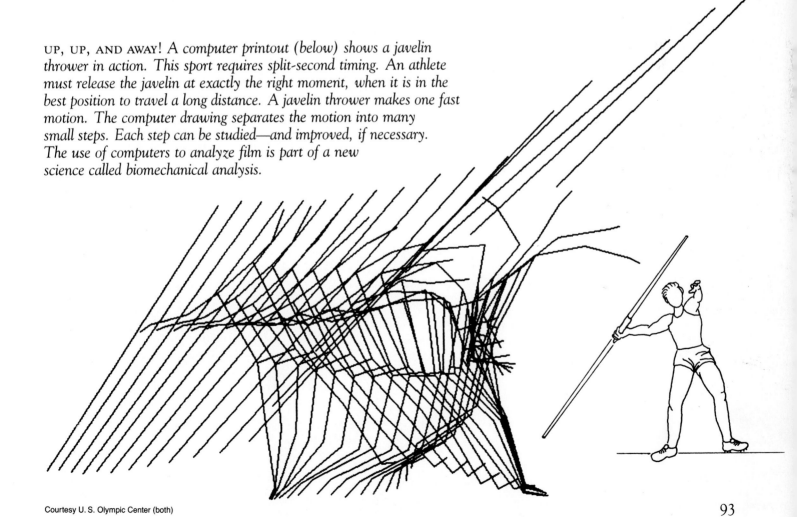

FLYING FLIP. *Bright lights and a camera catch circus acrobats in action. The woman springs off a tilt board, at the left, does a back flip, and lands at the right. The photographer's flash is focused on her, as is the spotlight. Her image on film is a strong one. The moving people on the ground are less brightly lit. Their images are weaker. Since the photographer's light flashes six times, the background is photographed six times. Each person in the picture is also photographed six times, producing many overlapping images. Some background shows through the weaker images, forming "ghosts."*
Harold E. Edgerton

LONG HOPS on *powerful hind legs earn the kangaroo rat its name. The rat pushes off with its hind legs. Then it pulls them up and forward for a landing. The entire jump covers about a foot (30 cm). A strobe light captures the action. Each flash of the light shows the rat at one point in its leap. The camera shutter stays open the entire time, so all of the images are on one piece of film.*

96

(Continued from page 92) spook house. Using his new device, Dr. Edgerton made pictures of things people had never seen before. He showed what a drop of milk looks like at the instant it hits a saucer full of milk. He even took pictures of bullets in flight!

Cameras can show us how our bodies move as we run, jump, and play. Computers can then translate the film into drawings. Using the drawings as a guide, athletes have been able to study their actions, correct their mistakes, and improve their skills. Scientists have learned more about animals by looking at pictures of how they move.

Special kinds of photography lead us into hidden worlds beyond the limits of our own vision. Now we can answer such questions as "How does an owl fly?" and "How does a blossom open?" We can slow motion down, and we can speed it up. With the help of film, a camera, and light, we can catch all the action.

Jane Burton, Bruce Coleman Ltd.

Stephen Dalton, Photo Researchers, Inc.

NIGHT FLIGHT *of an owl comes to life in a multiple-exposure photograph (above). This bird, which has the name "little owl," is about half the size of a pigeon.*

USING A LEAF *as a springboard, an insect called a rhododendron leafhopper takes off (left). A multiple-exposure photograph shows how the insect moves as it flies from plant to plant.*

GROWING UP. *A photographer made one picture of a plant every day for a week (right). She kept the camera in the same spot and used the same piece of film. At first, growth was slow. Then it speeded up. This small plant is a winter aconite (say* AK-uh-night). *It sometimes blooms in snowy weather.*

NATURE'S FIREWORKS. *Lightning seems to fill the sky above Kitt Peak National Observatory in Arizona. Several bolts of lightning were captured on one piece of film. The photographer left the camera shutter open for one minute. Each lightning flash gave off enough light to make an image on the film.*

98   Gary Ladd

WHO DISCOVERS HIDDEN WORLDS?

by Dr. Glenn O. Blough
Professor Emeritus, University of Maryland

N.G.S. Photographer Victor R. Boswell, Jr.

BRINGING SCIENCE TO LIFE *is the lifework of Dr. Glenn O. Blough, of Washington, D. C. He has taught science to elementary, secondary, and university students. He has written 35 books about science for young readers and their teachers. He was a professor of science education at the University of Maryland for 17 years and is a past president of the National Science Teachers Association. As a consultant and adviser, he now lends his experience and insight to Books for World Explorers.*

Once, not even the smartest people on earth knew the facts and figures you have learned from reading *Hidden Worlds*. They had no way of finding them out. There were no microscopes, telescopes, or photographic equipment to help them. Many people believed that the earth was flat and that it was the center of the universe. They thought everything was made of four elements: earth, air, fire, and water. They thought fierce dragons caused eclipses by swallowing the sun and the moon.

Most people accepted such explanations for the things they couldn't understand. But some people didn't accept everything they were told. They asked questions. They observed the world around them and tried to find out more about it.

One such questioner was an Italian professor of mathematics named Galileo. He didn't believe many of the old ideas, and he did experiments that proved some of them wrong.

In 1609, Galileo made one of the first telescopes. It magnified things only 3 times. But soon he built one that magnified objects 32 times. When Galileo turned his telescope to the heavens, he had some great surprises. He saw mountains and craters on the moon. He discovered four of Jupiter's moons. He noticed the rings of Saturn. And he found that the Milky Way was made of countless distant stars. No one had ever seen these things before!

Today Galileo is known as the father of modern science. Why? Because he didn't take things for granted. He asked questions. He tried things out, over and over again. He didn't believe a thing was true unless he had good evidence. With his telescopes and his experiments, Galileo changed the way people thought about the physical world and the universe.

In the mid-1600s, a Dutch lens maker named Anton van Leeuwenhoek made a simple microscope. He used it to look at a drop of water. There he saw tiny living creatures moving about. Some people had suspected that there were living things too small to see with the eye. The microscope showed them clearly for the first time. Many years later, scientists discovered that some of these things were bacteria that could cause diseases. Scientists began to study such bacteria to learn how to fight them.

The invention of the telescope and of the microscope opened many hidden worlds. Building on this knowledge, scientists have made better and better instruments. With their help, we can travel into space and bring back rocks from the moon. We can look through telescopes at distant galaxies, and study tiny creatures under powerful microscopes. We can see through solid objects, and measure forms of light that people once didn't know existed.

But instruments are only as good as the people who use them. If you're not a careful observer, even the best lenses and machines won't help you. How much of the world *you* see is up to *you*!

INDEX

Bold type refers to illustrations; regular type refers to text.

Words to Remember

This book contains words that may be new to you. Their meanings are explained below. Knowing what these words mean should help you understand other science books.

ATOM. The smallest unit of a substance.

COMPOUND EYE. An eye that has many lenses.

COMPOUND MICROSCOPE. An instrument that greatly magnifies small objects by using more than one lens with light directed through them.

CORNEA. The layer of the eye that protects the iris and the pupil.

ELECTROMAGNETIC SPECTRUM. The various kinds of light, both visible and invisible. The electromagnetic spectrum includes gamma rays and X rays, ultraviolet, visible light, infrared, and radio waves.

ELECTRON. A tiny particle that carries a negative electric charge.

FIELD OF VISION. The area seen at one instant without moving the eyes.

FLUORESCENCE. A visible glow that occurs when certain materials absorb light from another source, such as an ultraviolet lamp.

FOVEA. The part of the retina where light-sensitive cells are most tightly grouped. The fovea is responsible for our keenest vision.

HAND LENS. A single magnifying lens often set in a frame with a handle.

IRIS. The colored part of the eye. It regulates the amount of light that enters the eye.

LASER. An instrument that produces a very strong beam of light. In laser light, all of the waves have exactly the same wavelength.

LIGHT-YEAR. The distance a beam of light travels through airless space in one year, about 6 trillion miles (10 trillion km).

OPTIC NERVES. The nerves that carry visual signals from the eyes toward the brain.

OPTICAL FIBERS. Flexible threads of glass or plastic that carry light and images.

PHOSPHORESCENCE. A visible glow similar to fluorescence, but continuing after the light is turned off.

POLARIZING. Blocking some light waves so that all of the remaining waves vibrate in one direction.

PUPIL. The opening in the iris through which light enters the eye.

RADIANT ENERGY. The energy carried by light waves.

RADIOGRAPH. Picture made with X rays.

RETINA. The light-sensitive area lining the inside of the eyeball.

SCANNING ELECTRON MICROSCOPE. An instrument that uses electrons to form a magnified, three-dimensional image.

SCLERA. The white outer layer that protects most of the eyeball.

SONAR. An instrument that uses sound waves and their echoes to detect hidden objects, often underwater.

SONOGRAM. A picture made with sonar.

STROBOSCOPE. An electronic device that flashes a bright light at regular intervals.

THERMOGRAPH. An instrument that measures the heat given off by objects. It turns the measurements into visible signals that can be recorded on photographic film.

VISIBLE SPECTRUM. Light that can be seen by the human eye.

WAVELENGTH. The distance between one part of a wave, usually the top, and the same part of the next wave.

Additional Reading

1 **Overall:** Adler, Irving, *The Changing Tools of Science*, John Day, 1973. Anderson, Lucia, *The Smallest Life Around Us*, Crown, 1978. Keen, Martin, *The Microscope and What You See*, Grosset and Dunlap, 1974. Marten, Michael, et al., *Worlds Within Worlds*, Holt, Rinehart, Winston, 1977. Schwartz, Julius, *Magnify and Find Out Why*, McGraw-Hill, 1973. Shippen, Katherine B., *Men, Microscopes, and Living Things*, Viking, 1967. Toomer, Derek, and Alan Cane, *Invisible World*, Danbury, 1975. Villiard, Paul, *The Hidden World: The Story of Microscopic Life*, Four Winds, 1975. Wolberg, Barbara J., *Zooming In: Photographic Discoveries Under the Microscope*, Harcourt Brace Jovanovich, 1974.

Insects: Ada, Frank Graham, *An Audubon Reader: Bug Hunters*, Delacorte, 1978. Brian, M. V., *The World of an Ant Hill*, Faber and Faber, 1979. Fabre, Jean Henri, *Insects*, Scribner, 1979. Hughes, Jill, *A Closer Look at Bees and Wasps*, Watts, 1977. Rowland-Entwistle, Theodore, *Insect Life, The World You Never See*, Rand McNally, 1976. Sandved, Kjell B., and Michael Emsley, *Insect Magic*, Penguin, 1979.

Microbes: Lewis, Lucia, *The First Book of Microbes*, Watts, 1972. Nourse, Alan E., *The First Book of Viruses*, Watts, 1976. Patent, Dorothy Hinshaw, *Bacteria: How They Affect Other Living Things*, Holiday House, 1980.

Light microscope: Idyll, C. P., *Abyss: The Deep Sea and the Creatures that Live in It*, Crowell, 1976. Lucas, Joseph, and Pamela Critch, *Life in the Oceans*, Dutton, 1974. National Geographic magazine: "Teeming Life of a Pond," by William H. Amos, August 1970; "Blue Water Plankton," by William M. Hamner, October 1974; "Those Marvelous, Myriad Diatoms," by Richard B. Hoover, June 1979; "Rotifers: Nature's Water Purifiers," by John Walsh, February 1979. Patent, Dorothy, *Microscopic Animals and Plants*, Holiday House, 1974. Schwartz, George, *Life in a Drop of Water*, Natural History, 1970. Stonehouse, Bernard, *A Closer Look at Plant Life*, Gloster, 1977. Time-Life, *Aquatic Miniatures*, 1977.

Scanning electron microscope: Gardner, Robert, *This is the Way it Works*, Doubleday, 1980. Grillone, Lisa, and Joseph Gennaro, *Small Worlds Close Up*, Crown, 1978. National Geographic magazine: "At Home With the Bulldog Ant," by Robert F. Sisson, July 1974; "Electronic Voyage Through an Invisible World," by Kenneth Weaver, February 1977. Scharf, David, *Magnifications*, Schocken, 1977.

2 Bok, Bart J. and Priscilla F., *The Milky Way*, Harvard University, 1974. Branley, Franklyn M., *Eclipse: Darkness in Daytime*, Crowell, 1973. *Black Holes, White Dwarfs, and Super Stars*, Crowell, 1976. *Comets, Meteoroids, and Asteroids*, Crowell, 1974. Claiborne, Robert, *The Summer Stargazer: Astronomy for Absolute Beginners*, Coward, 1975. Freeman, Mae and Ira, *The Sun, the Moon, and the Stars*, Random House, 1979. Mitton, Jacqueline and Simon, *Concise Book of Astronomy*, Prentice-Hall, 1978. Murray, Bruce, *Flight to Mercury*, Columbia University, 1976. National Geographic magazine: "Mars: Our First Close Look," January 1977; "Voyager's Historic View of Earth and Moon," July 1978; "What Voyager Saw: Jupiter's Dazzling Realm," by Rick Gore, January 1980; "The Riddle of the Rings," by Rick Gore, July 1981; "When the Space Shuttle Finally Flies," by Rick Gore, March 1981; "How to Catch a Passing Comet," by Kenneth Weaver, January 1974. National Geographic Society: Friedman, Herbert, *The Amazing Universe*, 1975; Gallant, Roy A., *Picture Atlas of Our Universe*, 1980; Rey, H. A., *The Stars: A New Way to See Them*, Houghton Mifflin, 1970, and *Find the Constellations*, Houghton Mifflin, 1976. Sarnoff, Jane, *Space: A Fact and Riddle Book*, Scribner, 1978. Simon, Seymour, *Look to the Night Sky*, Viking, 1977, and *The Long View Into Space*, Crown, 1979. Smith, Norman, *Space: What's Out There*, Coward, McCann and Geoghegan, 1976. Zim, Herbert S., *The Universe*, Morrow, 1973, and *Stars: A Guide to the Constellations, Sun, Moon, Planets*, Golden, 1975.

3 Burstein, John, *Slim Goodbody: The Inside Story*, McGraw-Hill, 1977. Halacy, Daniel S., Jr., *X-Rays and Gamma Rays*, Holiday House, 1969. Levin, Edith, *The Penetrating Beam: Reflections on Light*, Rosen, 1979. National Geographic magazine: "Harnessing Light by a Thread," by Allen Boraiko and Fred Ward, October 1979; "Loch Ness," (sonar) by William S. Ellis, June 1977; "Eyes of Science," by Rick Gore, March 1978; "How We Found the Monitor," (sonar) by John G. Newton, January 1975. Wilson, Ron, *How the Body Works*, Larousse, 1978.

4 Frisby, John, *Seeing: Illusion, Brain, and Mind*, Oxford, 1979. Julesz, Bela, *Foundations of Cyclopean Perception*, The University of Chicago Press, 1971. Paraquin, Charles H., *Eye Teasers: Optical Illusion Puzzles*, Sterling, 1976. Perry, John, and Victor B. Scheffer, *The Seeing Eye*, Scribner, 1971. Simon, Seymour, *The Optical Illusion Book*, Four Winds, 1976. Sislowitz, Marcel J., *Look! How Your Eyes See*, Coward, 1977.

5 Branley, Franklyn M., *The Electromagnetic Spectrum: Key to the Universe*, Crowell, 1979. Freeman, Ira M., *Science of Light and Radiation*, Random House, 1968. Horsburgh, Peg, *Living Light: Exploring Bioluminescence*, Messner, 1978. Kettlekamp, Larry, *Lasers: The Miracle Light*, Morrow, 1979. Morris, Richard, *Light*, Bobbs-Merrill, 1979. Mueller, Conrad G., and Mae Rudolph, *Light and Vision*, Time, Inc. Life Science Library, 1966. National Geographic magazine: "The Laser's Bright Magic," by Thomas Meloy, December 1966; "Remote Sensing: New Eyes to See the World," by Kenneth F. Weaver, January 1969. *Science Projects: Mirrors*, Booklab, 1972. *Science Puzzles Pictures*, Booklab, 1972. Schneider, Herman, *Laser Light*, McGraw-Hill, 1978. *Science Fun With a Flashlight*, McGraw-Hill, 1975. *Science Fun for You in a Minute or Two*, McGraw-Hill, 1975.

6 Andersen, Yvonne, *Make Your Own Animated Movies*, Little, Brown, 1970. Bostrum, Roald, *A Look Inside Cameras*, Raintree, 1981. Jacobs, Lou, Jr., *You and Your Camera*, Lothrop, Lee, and Shepard, 1971. Johnson, James P., *Photography for Young People*, McKay, 1971. Laycock, George, *A Complete Beginner's Guide to Photography*, Doubleday, 1979. Palder, Edward L., *Magic With Photography*, Grosset and Dunlap, 1969. Sandler, Martin W., *The Story of American Photography: An Illustrated History for Young People*, Little, Brown, 1979. Time-Life, *Photography as a Tool*, Time-Life Books, 1970.

Poster: Adams, Richard, *Nature Through the Seasons*, Simon and Schuster, 1975. Field Enterprises, *The Green Kingdom*, 1974. Lexau, Joan M., *The Spider Makes a Web*, Hastings House, 1979. Newcomb, Lawrence, *Newcomb's Wildflower Guide*, Little, Brown, 1977. Reader's Digest, *Joy of Nature*, 1977.

Consultants

The Special Publications and School Services Division is grateful to the individuals, organizations, and agencies named or quoted in the text and to the individuals cited here for their generous assistance: Dr. Alice Alldredge, *University of California at Santa Barbara*; Lila Bishop, *Sidwell Friends School*; Dr. Glenn O. Blough, *Educational Consultant*; Donald F. Brandewie, *Arlington County Schools*; Edward C. Collins, *The Perkin-Elmer Corporation*; Ronald I. Crombie, *Smithsonian Institution*; Elizabeth C. Dudley, *University of Maryland*; John E. Fletcher; Dr. Theodore R. Gull, *Goddard Space Flight Center*; Sarah H. Haigler; Gary F. Hevel, *Smithsonian Institution*; Judith M. Hobart, *Beauvoir School*; Maris Juberts, *National Bureau of Standards*; Patricia Leadbetter King, *National Cathedral School for Girls*; Dr. Marilyn J. Koering, *George Washington University Medical Center*; Dr. Nicholas J. Long, *Consulting Psychologist*; Tom McIntyre, *National Marine Fisheries Service*; Dr. Charles Montrose, *The Catholic University of America*; Dr. David A. Newsome, *National Eye Institute*; Roger Ratcliffe, *Department of Agriculture*; Dr. George E. Watson, *Smithsonian Institution*; Dr. Austin B. Williams, *Smithsonian Institution*.

Composition for *Hidden Worlds* by National Geographic's Photographic Services, Carl M. Shrader, Chief; Lawrence F. Ludwig, Assistant Chief. Printed and bound by Holladay-Tyler Printing Corp., Rockville, Md. Color separations by the Beck Engraving Company, Philadelphia, Pa.; the Lanman-Progressive Corp., Washington, D. C.; Lincoln Graphics, Inc., Cherry Hill, N.J.; NEC, Inc., Nashville, Tenn.

Library of Congress CIP Data

Main entry under title:

Hidden worlds.

(Books for world explorers)

Bibliography: p.
Includes index.
Contents: Think small / by Edith Kay Pendleton—Beyond earth / by Kathryn Allen Goldner and Carole Garbuny Vogel—The inside story / by Catherine O'Neill—[etc.]
1. Vision—Juvenile literature. 2. Microscope and microscopy—Juvenile literature. 3. Telescope—Juvenile literature. 4. Radiography—Juvenile literature. [1. Vision. 2. Microscope and microscopy. 3. Telescope. 4. Radiography] I. National Geographic Society (U. S.) II. Series.

QP475.7.H52   500   79-3244
ISBN 0-87044-336-4 (regular binding)   AACR2
ISBN 0-87044-341-0 (library binding)

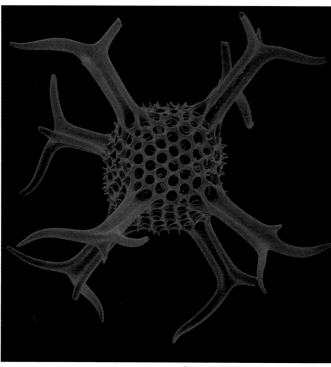

Cover: Manfred Kage from Peter Arnold, Inc.

COVER: *This skeleton belongs to a sea creature no larger than the point of a sharp pencil. It is a radiolarian (say ray-dee-oh-LAR-ee-un). When it was alive, it had a soft body inside the round part of its skeleton. Hundreds of kinds of radiolarians drift near the surface of the oceans. All have colorless skeletons. Color was added to this one.*

HIDDEN WORLDS

PUBLISHED BY
THE NATIONAL GEOGRAPHIC SOCIETY
WASHINGTON, D. C.

Gilbert M. Grosvenor, *President*
Melvin M. Payne, *Chairman of the Board*
Owen R. Anderson, *Executive Vice President*
Robert L. Breeden, *Vice President,
Publications and Educational Media*

PREPARED BY THE SPECIAL PUBLICATIONS
AND SCHOOL SERVICES DIVISION

Donald J. Crump, *Director*
Philip B. Silcott, *Associate Director*
William L. Allen, William R. Gray, *Assistant Directors*

STAFF FOR BOOKS FOR WORLD EXPLORERS: Ralph Gray, *Editor*; Pat Robbins, *Managing Editor*; Ursula Perrin Vosseler, *Art Director*

STAFF FOR HIDDEN WORLDS:
Ross Bankson, Pat Robbins, *Text Editors*
Alison Wilbur, *Picture Editor*
Beth Molloy, *Designer*
Mary B. Campbell, Tee Loftin,
Nancy J. Watson, *Researchers*
Mary Elizabeth Davis, *Editorial Assistant*
Artemis S. Lampathakis, *Illustrations Secretary*

FAR-OUT FUN! AND SUPPLEMENTARY ACTIVITIES: Patricia N. Holland, *Project Editor*; Roger B. Hirschland, *Text Editor*; Roz Schanzer (pages 3, 10, 17); Marsha Lederman (pages 4-5, 15, 16); Mary Dustin (pages 6, 7); Barbara Gibson (pages 20, 23 bottom, 24), *Artists*; Art Iddings (pages 2, 8-9, 18-19, 22-23), *Mechanicals*.

ENGRAVING, PRINTING, AND PRODUCT MANUFACTURE: Robert W. Messer, *Manager*; George V. White, *Production Manager*; Gregory Storer, *Production Project Manager*; Mark R. Dunlevy, Richard A. McClure, Raja D. Murshed, Christine A. Roberts, David V. Showers, *Assistant Production Managers*; Susan M. Oehler, *Production Staff Assistant*

STAFF ASSISTANTS: Debra A. Antonini, Nancy F. Berry, Pamela A. Black, Barbara Bricks, Nettie Burke, Jane H. Buxton, Claire M. Doig, Rosamund Garner, Victoria D. Garrett, Nancy J. Harvey, Joan Hurst, Suzanne J. Jacobson, Virginia A. McCoy, Merrick P. Murdock, Cleo Petroff, Victoria I. Piscopo, Jane F. Ray, Carol A. Rocheleau, Katheryn M. Slocum, Jenny Takacs, Phyllis C. Watt

MARKET RESEARCH: Joe Fowler, Carrla L. Holmes, Meg McElligott, Stephen F. Moss, Marjorie E. Smith, Susan D. Snell

INDEX: Colleen B. DiPaul